13th International Colloquium on Atomic Spectra and Oscillator Strengths for Astrophysical and Laboratory Plasmas

13th International Colloquium on Atomic Spectra and Oscillator Strengths for Astrophysical and Laboratory Plasmas

Editors

Tomas Brage
Roger Hutton
Jun Xiao

MDPI • Basel • Beijing • Wuhan • Barcelona • Belgrade • Manchester • Tokyo • Cluj • Tianjin

Editors

Tomas Brage
Lund University
Sweden

Roger Hutton
Fudan University
China

Jun Xiao
Fudan University
China

Editorial Office
MDPI
St. Alban-Anlage 66
4052 Basel, Switzerland

This is a reprint of articles from the Special Issue published online in the open access journal *Atoms* (ISSN 2218-2004) (available at: https://www.mdpi.com/journal/atoms/special_issues/ASOS2019).

For citation purposes, cite each article independently as indicated on the article page online and as indicated below:

LastName, A.A.; LastName, B.B.; LastName, C.C. Article Title. *Journal Name* **Year**, *Article Number*, *Page Range*.

ISBN 978-3-03943-146-5 (Hbk)
ISBN 978-3-03943-147-2 (PDF)

Contents

About the Editors

Tomas Brage (PhD) is a professor at Lund University. He obtained his PhD in Atomic Physics in 1988, and has, since then, had positions as a research assistant professor of Computer Science at Vanderbilt University and as a research associate at the NASA Goddard Space Flight Center, working on the Goddard High Resolution Spectroscopy Mission on the Hubble Space Telescope. For the last twenty years, he has been a Professor of Physics at the Division of Mathematical Physics, the Department of Physics, Lund University in Sweden, and until 2018, the Director of Education in Physics at the Faculty of Science. His main research interests are laboratory astrophysics and computational atomic physics, and he has published over 100 articles in refereed journals. Until 2019, he was a visiting professor at the Fudan University in Shanghai, China, where he spends a few months each year at the Institute of Modern Physics. For the last 15 years, he has been strongly involved in work on gender and science, where he is active in several European networks, e.g. as a steering group member of the policy group for equality, diversity and inclusion of the League of European Research Universities (LERU) network, and a work-group leader for gender-dimension in physics of the GENERA network. He has served as an expert for several Horizion2020 projects and is currently a member of the GenderEX project. He has led several projects at Lund, e.g. the Gender Certification project, Antidiscrimination education and, most recently, the Core-Value project.

Roger Hutton (PhD) is a professor at Fudan University. Professor Hutton obtained his PhD in Atomic Physics from the University of Lund in 1988. Following his PhD, he spent two years as a postdoctoral researcher at the Lawrence Berkeley and Livermore National laboratories, working mainly on low energy ion–atom collisions. This was followed by a third year as a post doc at the Manne Siegbahn Laboratory in Stockholm, as an assistant to Prof. Sheldon Datz, who was the Tage Elander Professor at the time. In 1991, Prof. Hutton returned to Lund as a research assistant in atomic spectroscopy. He stayed in Lund until 2005, when he became a professor at the Institute of Modern Physics, Fudan University. During the period 1998–2000, he was a visiting researcher at the Atomic Physics laboratory at the RIKEN laboratory in Japan. His main line of research is atomic spectroscopy, with applications in both fusion and astrophysical plasmas. He has penned over 100 research papers in refereed journals and co-chaired/hosted many international meetings, notably the PEARL series, which was mainly held in China between 2005 and 2016.

Jun Xiao (PhD) is an associate professor at the Institute of Modern Physics, Fudan University. His research interests include atomic physics, the spectroscopy of highly charged ions, physical processes in fusion and astrophysical plasmas, the development of EBIT device, etc. In 2011, Dr. Xiao received his PhD in atomic physics from Fudan University. He developed an ultra-low-energy EBIT, namely SH-HtscEBIT, in 2012, and has published over 30 articles in refereed journals. He became a Scientific Organizing Committee member of the ASOS meeting in 2019.

Editorial

The 13th International Colloquium on Atomic Spectra and Oscillator Strengths for Astrophysical and Laboratory Plasmas

Jun Xiao [1],*, Tomas Brage [2] and Roger Hutton [1]

[1] Shanghai EBIT Laboratory, Key Laboratory of Nuclear Physics and Ion-Beam Application (MOE), Institute of Modern Physics, Fudan University, Shanghai 200433, China; rhutton@fudan.edu.cn

[2] Division of Mathematical Physics, Department of Physics, Lund University, 221 00 Lund, Sweden; tomas.brage@teorfys.lu.se

* Correspondence: xiao_jun@fudan.edu.cn

Received: 4 August 2020; Accepted: 7 August 2020; Published: 11 August 2020

The 13th International Colloquium on Atomic Spectra and Oscillator Strengths for Astrophysical and Laboratory Plasmas (ASOS2019), co-hosted by Fudan and Lund Universities, was held at Fudan University from 23–27 June 2019. It attracted over 100 participants, both from outside China, with about 40 participants from Argentina, Belgium, France, Germany, Japan, Lithuania, The Netherlands, Poland, Russia, South Korea, Sweden, the United Kingdom, and the United States, but also over 60 scholars from universities or research institutes from within China (e.g., the Chinese Academy of Sciences, China Academy of Engineering Physics, the Southwestern Institute of Physics, Fudan University, Northwest Normal University, Lanzhou University, National University of Defense Technology, University of Science and Technology of China, Shaanxi Normal University, Tsinghua University, and Shanghai Jiao Tong University). Establishing itself in PRC, the global network supported by the conference is thereby extended. The ASOS-conferences started in 1983, with what was basically a trial meeting held at the University of Lund. Sadly both Professors Indrek Martinson and Svenneric Johansson, instrumental in the first ASOS, are no longer with us but are certainly remembered by many in the ASOS community. On the other hand, two of the current co-chairs of the ASOS-3 SOC both attended the very first meeting as graduate students in Lund, showing it is a small world. We are sure that all the participants of ASOS-13 will remember the wonderful organisation and for that we would like thank the members of the local organising committee, led by Professor Baoren Wei and Associate Professor Jun Xiao (Figures 1 and 2).

Figure 1. Prof. Tomas Brage (**left**) and Assoc. Prof. Jun Xiao giving the opening speech of ASOS2019.

Figure 2. Prof. Roger Hutton (**left**) and Prof. Tomas Brage continue the opening speech and welcome to the participants.

Dedicated Session to Charlotte Froese Fischer

Professor Alan Hibbert from the Queen's University Belfast delivered a report entitled "Charlotte Froese Fischer: Her works and her impact", discussing the career of one of the corner stone figures in the field of atomic structure and spectroscopy, a way to celebrate her 90th birthday. Alan's talk was showing respect for her tireless scientific work for more than 60 years and her ground-breaking scientific contributions. It was a pity that Charlotte was not able to participate in the conference, but as a pleasant surprise, Alan showed a recorded message from Charlotte for the participants, which added a special commemorative significance to this colloquium. As part of our theme to celebrate Charlotte's 90th birthday a number of contributions were given by some of her more recent post docs, from Gediminas Gaigalas in the 1990s to Ran Si in last two years (Figures 3 and 4).

Figure 3. Prof. Alan Hibbert (**left**) giving a report and showing a video message from Prof. Charlotte Froese Fischer.

Figure 4. Drs. Ran Si (**left**) and Philip Judge giving talks.

A total of 38 experts and scholars were invited to give lectures during the four-day seminar and the results of their endeavours can be found at https://www.koushare.com/video/meetingVideo?mid=89. The contents covered a number of major themes within atomic physics, both theoretical and experimental, including applications to, e.g., astrophysics and fusion plasma. The meeting spanned over four full days. Since it was held in Shanghai, with excellent possibilities to explore the enormous city, the participants could enjoy their own sight seeings and hence no joint outing was organized. The conference dinner was held at a rotating restaurant in the center of Shanghai, with a great view of the city, from all angles.

We are now looking forward to the 14th ASOS-meeting, which is planned for Paris in 2022—until then, stay safe! (Figure 5).

Figure 5. The conference photo of ASOS2019.

Funding: This research received no external funding.

Conflicts of Interest: The authors declare no conflict of interest.

Article

Orthogonal Operators: Applications, Origin and Outlook

Peter Uylings [1,*] and Ton Raassen [1,2]

[1] Anton Pannekoek Institute for Astronomy, University of Amsterdam, Science Park 904,
     1098 XH Amsterdam, The Netherlands; a.j.j.raassen@sron.nl
[2] SRON Netherlands Institute for Space Research, Sorbonnelaan 2, 3584 CA Utrecht, The Netherlands
*   Correspondence: p.uylings@contact.uva.nl

Received: 19 October 2019; Accepted: 8 November 2019; Published: 13 November 2019

Abstract: Orthogonal operators can successfully be used to calculate eigenvalues and eigenvector compositions in complex spectra. Orthogonality ensures least correlation between the operators and thereby more stability in the fit, even for small interactions. The resulting eigenvectors are used to transform the pure transition matrix into realistic intermediate coupling transition probabilities. Calculated transition probabilities for close lying levels illustrate the power of the complete orthogonal operator approach.

Keywords: atomic lifetime and oscillator strength determination; theoretical modeling and computational approaches; atomic databases and related topics

1. Introduction

Since its first introduction [1], the orthogonal operator technique has appeared to be a powerful tool in reducing the deviations between calculated and experimental energy values in complex spectra ($Z > 20$). Due to its orthogonality, the operator set is stable enough to introduce small (thus far neglected) higher-order magnetic and electrostatic effects in the fitting procedure. By this extension, deviations between calculated and experimental energy values frequently approach experimental accuracy [2]. More recent use is found in the FERRUM project [3] and in the analysis of 5d-spectra [4]. Origin and subsequent developments underlying the method are discussed. The operator inner product is shown to be a property of operators rather than of its accidental matrix elements by a general progression formula as a function of the number of electrons. Linear algebra can now fruitfully be used to project a variety of contributions onto the orthogonal operator set, both analytically and numerically. The actual construction of an orthogonal operator set is illustrated for $d^n p$ configurations. Also, the orthogonal operator method is positioned (as to overlap and differences) with respect to other approaches for describing complex spectra, such as large-scale use of Cowan's code or Multi-Configuration Dirac-Hartree-Fock (MCDHF) calculations. The description of the odd and even spectra of Fe VI are used as a running example, but other regions of application are mentioned. *Ab initio* calculations as well as conversion of operator sets are considered, and the interplay between explicit and implicit configuration interaction is discussed. Possibly controversial issues such as (over)completeness, term dependency and a truncation of the model space are reviewed. The accurate description of the energy structure is expected to result in optimally calculated eigenvector compositions. Naturally, this property can be exploited to calculate accurate electric dipole (E1), magnetic dipole (M1) and electric quadrupole (E2) transition probabilities. How polarization, core and valence excitations and full relativity are presently implemented and can be improved in the near future, will be discussed. We recently reinstalled our original database with transition arrays of the 3d and 5d shell [5]. We intend to cooperate with other groups and increase the accessibility of the method.

2. Applications

Why should one use orthogonal operators when a conventional Slater–Condon approach such as Cowan's code [6,7] is so easy-to-use as a universal tool? Cowan's code has been a blessing for the experimental atomic physicist for the last 50 years, and it will no doubt continue to be so for many years to come. The orthogonal operator approach may be considered as both an extension and a refinement of the conventional least-squares fitting (LSF) approach, but it also raises the need of finding physically acceptable initial estimates for quite a number of small parameters, especially with an incomplete spectrum. The stability of a parameter versus change or addition of others, on the other hand, is a great asset of the method. Figure 1 provides a clear picture of this aspect [8].

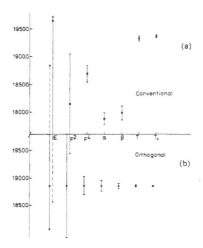

Figure 1. (**a**) The values of the E_{av} parameter in a number of LSF to the $3d^3$ configuration in Cr IV. In the first fit (from left to right) only the E_{av} parameter was allowed to vary with all other parameters fixed at zero while in subsequent fits the parameter indicated on the abscissa was added. (**b**) The values of the E_{av} parameter using the same procedure as in (**a**), but using a set of orthogonal equivalents.

Both the conventional and the orthogonal operator approaches are based on semiempirical LSF of physical parameters. The mean error can be interpreted as the 'blobsize' used by painters (Seurat and others) of the art-movement 'Pointillism' [9]. They used blobs for their paintings and could not represent details smaller than the blobsize. In the same way we cannot describe level structures in detail when the spacing between levels is smaller or comparable to the mean error.

In quite a number of cases, the mean error of an orthogonal operator LSF is smaller by an order of magnitude w.r.t. a conventional LSF. In principle, this leads to better eigenvector compositions and thus better transition probabilities. In some cases, refinements are expedient and they will be discussed below. In Table 1, some of the characteristics of both methods are set side by side.

The term analysis of complex spectra like Mn IV [10], Re III [11] and Os III [12] only came within reach by the accuracy of orthogonal operator predictions. Such predictions concern both energy levels and electric dipole intensities.

In addition, reliable calculations of forbidden magnetic dipole and electric quadrupole intensities are enabled by the accurate eigenvectors from an orthogonal operator LSF of the pertinent system. On a more theoretical level, use of orthogonal operators allows establishing small and thus far neglected interactions that reflect many-body, relativistic and correlation effects. Fully relativistic *ab initio* perturbation or Configuration Interaction (CI) calculations can subsequently be compared to the parameter values found.

Table 1. Characteristics of the conventional and the orthogonal operator LSF methods.

	Conventional	Orthogonal Operators
# configurations	large	limited
parameter interdependence	yes	no
2-body electrostatic	yes	yes
3-body electrostatic	only T_1 and T_2	yes
4-body electrostatic	no	yes
1-body magnetic	yes	yes
2-body magnetic	no	yes
mean error	medium	small
initial preparation	small	medium
transition probabilities	generally sufficient	close to experiment
use	automated	more case to case

2.1. Oscillator Strengths Involving Close Lying Levels

To illustrate the importance of accurate eigenvectors, two examples in the spectra of Fe III and Fe II are given below. The effect is especially striking for close lying levels, when the actual separation may be smaller than the mean error of the fit: the 'blobsize' is here too large for a reliable calculation.

First, the eigenvectors of the conventional and the orthogonal method in the Fe III case are compared in Table 2 [13]:

Table 2. Two close lying levels in the $J = 5$ matrix of the $3d^5 4p$ configuration of Fe III calculated by the conventional and the orthogonal method.

Conventional Method, Overall Mean Error $\sigma = 139$ cm^{-1}				
Exp	Calc	Diff.	\multicolumn{2}{c}{Eigenvector composition}
139509.2	139407.4	101.8	49%	$(^2H)^3I + 21\%(^4F)^3G$
139463.0	139378.4	84.7	49%	$(^2H)^3I + 32\%(^4F)^3G$

Orthogonal method, overall mean error $\sigma = 12$ cm^{-1}				
Exp	Calc	Diff.	\multicolumn{2}{c}{Eigenvector composition}
139509.2	139504.1	5.0	83%	$(^2H)^3I + 6\%(^2H)^3H$
139463.0	139476.0	−13.0	44%	$(^4F)^3G + 29\%(^2G)^3G$

With deviations lower by an order of magnitude, it is seen that the eigenvector composition changes appreciably. Next, the corresponding transition probabilities (gA-values) are compared to experiment in Table 3:

Table 3. Transition probabilities (gA) in Fe III calculated by the conventional (Conv.) and the orthogonal (Orth.) method, compared to experiment [14]: B. are estimates of photographic blackening on a logarithmic scale, Int. are scaled intensities calculated from B.

The $3d^5 4s$-$3d^5 4p$ Array					
λ (Å)	B.	Int.	Conv. (10^8 s^{-1})	Orth. (10^8 s^{-1})	Transition
2041.203	14	22.3	18.3	26.3	$(^4F)^3F_4 - 139463.0$
2039.283	11	2.8	17.2	4.40	$(^4F)^3F_4 - 139509.2$
2012.901	10	1.4	6.95	2.58	$(^2G)^3G_4 - 139463.0$
2011.034	13	11.2	6.07	11.0	$(^2G)^3G_4 - 139509.2$

The conventional method calculates two equally strong lines and then two equally weaker lines, while the experiment shows a strong line, a weaker line, a little bit weaker line and then again a stronger line but not as strong as the first line. This pattern is exactly described by the orthogonal operator approach.

Another striking example of the impact of the mixing percentages, i.e. the eigenvector accuracy, on the oscillator strengths is given by Hibbert [15,16] in the spectrum of Fe II. Two lines at 2507 Å and 2509 Å originating from the same (even) lower level $^4F_{7/2}$ are considered; the two close (odd) upper levels are given in Table 4:

Table 4. Mixings in $J = 9/2$ levels of Fe II.

		Percentages		
	Level↓	$(4p)^4G$	$(5p)^6F$	$(5p)^4F$
Raassen and Uylings [17]	90040.5	16	36	41
	90072.7	76	9	8
Corrégé and Hibbert [15]	90042.8	43	29	13
	90067.4	41	27	16
Uylings and Raassen *	90042.7	44	26	22
	90067.4	48	23	22

* This work.

In Table 5, the eigenvector percentages are from our original calculation [17], a calculation of Corrégé and Hibbert [15] fine-tuned to the experimental energy levels, and our present contribution with a similar manipulation of the E_{av} parameter. The effect of the fine-tuning is obvious: the level percentages are seen to approach each other very closely and turn out to be quite sensitive to the eigenvalue calculation, even at the sub- cm^{-1} level.

The corresponding oscillator strengths are given in Table 5, together with the result of a conventional calculation [18]. As to be expected, the oscillator strengths are accordingly sensitive to the eigenvector composition, and finally turn out to become about equal. This result is consistent with an earlier successful modeling of the emission spectrum of η Carinae [19].

Table 5. Oscillator strengths of the 2507, 2509 Å lines in Fe II.

Source	2507.552	2509.097	Sum
Kurucz [18]	0.001	0.297	0.298
Raassen and Uylings [17]	0.237	0.045	0.282
Corrégé and Hibbert [15]	0.138	0.136	0.274
Uylings and Raassen*	0.148	0.134	0.282

* This work.

2.2. Interplay with ab initio Calculations

The inner product of two operators t and u is defined as:

$$t : u = \sum_{\Psi, \Psi'} \langle \Psi | t | \Psi' \rangle \langle \Psi' | u | \Psi \rangle \tag{1a}$$

where the summation runs over all states Ψ, Ψ' of the system; it is basically the (double) contraction of the two matrices. The inner product is commutative by definition.

Taking magnetic degeneracy into account, it can be reduced to:

$$t : u = \sum_{\Psi_J, \Psi'_J} (2J + 1) \langle \Psi_J | t | \Psi'_J \rangle \langle \Psi'_J | u | \Psi_J \rangle \tag{1b}$$

or, in the case of electrostatic operators:

$$t : u = \sum_{\Psi_{SL}, \Psi'_{SL}} (2S+1)(2L+1) \left\langle \Psi_{SL} \mid t \mid \Psi'_{SL} \right\rangle \left\langle \Psi'_{SL} \mid u \mid \Psi_{SL} \right\rangle \tag{1c}$$

Equating $t : u$ to zero is used to define a set of orthogonal operators to be used in a LSF procedure: as a result, each parameter is now equipped with its own unique 'behavior'. However, use of the concept of operator inner product does not stop there. The resulting linear algebra allows the definition of an operator projection and this opens up new possibilities that can fruitfully be exploited in *ab initio* calculations. Let the operator space be spanned by a set of orthogonal operators $H_i = p_i P_i$ with $p_i : p_j = 0 \ (i \neq j)$, where the angular operators form an orthogonal set $\{p_i\}$ and the radial factors P_i are treated as parameters. Any arbitrary operator $U = vV$ can now be expressed in terms of the complete basis set of orthogonal operators:

$$v = \sum_i \alpha_i \cdot p_i = \sum_i \frac{v : p_i}{p_i : p_i} \cdot p_i \tag{2}$$

The expression for the individual contribution ΔP_i of a single operator U to a parameter P_i and (when summed over all possible contributing operators U) its complete *ab initio* calculation follows:

$$\Delta P_i = \alpha_i \cdot V = \frac{v : p_i}{p_i : p_i} \cdot V \rightarrow P_i = \sum_{U=vV} \frac{v : p_i}{p_i : p_i} \cdot V \tag{3}$$

Moreover, the projection of any physical operator $U = vV$ on a finite (and possibly incomplete) basis $\{p_i\}$ is complete if and only if the magnitude of the operator equals the sum of the magnitudes of its projections:

$$v : v = \sum_i \frac{(v : p_i)^2}{p_i : p_i} \tag{4}$$

The simple projection formula (3) can be used to derive analytical expressions, e.g. for the contribution of a variety of two-particle effects, relativistic and perturbative, to the spin-orbit parameter ζ [20]. Analytic expressions for the important [4] orthogonal $d^n s$ operators T_{dds} (3-body electrostatic) and A_{mso} (2-body magnetic) were derived in this way as well [2,21].

Alternatively, equation (3) is readily programmed to allow numerical projections, e.g., of Slater integrals or perturbative effects onto the orthogonal operator set to calculate the contributions to any parameter of your choice.

Earlier *ab initio* calculations of parameters, whether analytical or numerical, could only take into account contributions that are directly proportional to the operator concerned. Equation (3), on the other hand, allows us to calculate all *ab initio* contributions with a non-zero inner product.

This is exemplified in Table 6 by an *ab initio* calculation of the Trees operator T_1 in Fe VI ($3d^3$). Nominally, T_1 'only' accounts for $s \rightarrow 3d$ and $3d \rightarrow s$ excitations. Using Equation (3), one finds that there are also non-zero contributions from $3d \rightarrow d', g, i$ excitations with a different spin-angular character. A second example of this method is the *ab initio* calculation of T_{dds} and A_{mso} in Fe VI ($3d^2 4s$), given in Table 7.

Another fruitful strategy is fitting to MCDHF calculated energy levels (single configuration). Keeping the parameters associated with effective operators fixed to zero, the mean error of such fits is generally well below 1 cm^{-1}. This demonstrates the completeness of the orthogonal operator description in the case of full relativity. The resulting parameter values may be compared to a direct *ab initio* calculation from wavefunction integration. The results turn out to be completely equivalent, as illustrated by the below example of Fe VI ($3d^3$).

Table 6. Second-order contributions $\Delta T_1 \propto R^k R^{k'}/\Delta E$ to the three-particle Trees parameter in Fe VI ($3d^3$).

Exc. (kk')	22	24	44
$s \to 3d$	−12.067	-	-
$3d \to s$	0.209	-	-
$3d \to d'$	−0.198	0.405	−0.129
$3d \to g$	2.391	0.710	−1.107
$3d \to i$	-	-	0.037
Total calc.		−9.727	
Fitted value		−8.452	

Table 7. Calculated contributions to T_{dds} and A_{mso} in Fe VI ($3d^2 4s$) compared with experiment.

	T_{dds}		A_{mso}
$3d \to d'$	27.8	$3d \to d'$	1.60
$4s \to d'$	−118.9	$-\frac{6}{5} \cdot W^1$	−0.18
$4s \to g$	3.0	$4 \cdot N^0$	1.78
Total calc.	−88.1		3.20
Fitted value	−91.2		2.98

The general relativistic form of the traditional Slater integral [21]:

$$R^k(ab,cd) = \frac{1}{4} \sum_{j_a,j_b,j_c,j_d} [j_a,j_b,j_c,j_d] \begin{Bmatrix} j_a & k & j_c \\ l_c & \frac{1}{2} & l_a \end{Bmatrix}^2 \begin{Bmatrix} j_b & k & j_d \\ l_d & \frac{1}{2} & l_b \end{Bmatrix}^2$$
$$\times \int_0^\infty \int_0^\infty dr_1 dr_2 \, (F_a F_c + G_a G_c)_1 \, r_<^k/r_>^{k+1} \, (F_b F_d + G_b G_d)_2 \tag{5}$$

is straightforwardly integrated with the MCDHF wavefunctions of Fe VI ($3d^3$) to yield: $F^2(3d,3d) = 112493.69$. On the other hand, a fit of orthogonal operators to the corresponding Dirac energy levels gives: $O_2 = 12312.72$ and $O_2' = 8814.92$, from which $F^2(3d,3d) = \frac{9}{20} \cdot \sqrt{140} \cdot (O_2 + O_2') = 112493.53$. The closeness of the results demonstrates the ability of the orthogonal operator method to retrieve the correct physical information from the data.

This property may also be used to obtain initial values for two-body magnetic parameters. Two-body magnetic parameters describe mutual spin-orbit (MSO) and electrostatic spin-orbit (EL-SO) interactions. These interactions are both included in a MCDHF calculation: the first in the (Dirac-)Breit interaction and the second by the fact that single electron excitations of the type $nl \to n'l$ are to a large extent included in the Hartree-Fock potential (Brillouin's theorem) [6]. In addition, there are two-body magnetic operators for spin-spin effects (included in the Breit interaction) as well.

The impact of the values of the one- and two-body magnetic parameters on the mean error σ of the fit is shown in Table 8 for Fe VI ($3d^3$).

The operators associated with A_c to A_6 have spin-orbit character while A_1 and A_2 describe spin-spin effects. In each case, all other (electrostatic) parameters were left free to vary. In the column Fit(1), the two-body parameters A_i were fixed to zero. The A_i values of a pure Dirac-Fock (DF) calculation only deteriorate the fit: it can be concluded that addition of the Breit interaction (DF+Breit) is essential for improvement; the DF and DF+Breit calculations are carried out with the GRASP92 package [22].

In the column headed B-splines, a complete first and second order Hartree-Fock (HF) calculation of ζ_d and the A_i parameters is carried out; the channel for the $3d \to nd$ excitations is calculated with a B-splines program developed in Amsterdam [23,24]. This program employs the effective completeness of B-splines to calculate channels of one-electron excitations from a frozen HF core potential. All values can be compared with the results of the final fit Fit(2).

Table 8. Values of one- and two-body magnetic operators in Fe VI $3d^3$.

	Fit(1)	DF	DF + Breit	B-splines	Fit(2)
ζ_d	578.63	636.97	579.56	598.09	594.52
A_C	0	4.43	2.95	3.16	2.84
A_3	0	0.18	2.07	1.97	2.41
A_4	0	4.39	4.37	4.31	3.86
A_5	0	1.64	7.18	7.05	6.86
A_6	0	2.33	−9.22	−9.08	−9.85
A_1	0	−0.12	0.41	0.88	0.90
A_2	0	0.12	−2.31	−2.73	−2.90
σ	28.3	73.4	14.2	5.8	1.9

2.3. Configuration Interaction

It is well known that the central field model is flawed by configuration interaction (CI). This is particularly relevant in the spectra of doubly ionized, singly ionized and neutral atoms, where the single configuration model is increasingly breaking down. For the iron group elements, this is mainly due to energetically favorable $3d \rightarrow 4s$ substitutions. More specifically, effective parameters that account for weak configuration interactions with large numbers of high-lying configurations, are not able to do a reliable job in the spectra with a lower degree of ionization. Higher order electrostatic and magnetic effects (described by effective orthogonal operators) can only be determined reliably if the first order model is reasonably accurate. Evidently, the model space has to be expanded to include nearby configurations. Yet, in the balance between (perturbative) effective parameters and (variational) explicit interactions, with orthogonal operators we like to retain the first and limit the second to the ones most necessary. Still, in Fe II, this implies already at least 6 configurations of the same parity for the lower even and odd systems.

The orthogonal operator approach is at its best for operators with many diagonal matrix elements, so the inherent off-diagonal character of CI admittedly reduces the strength of the approach. To obtain consistent iso-electronic parameter extrapolations, it is necessary to keep the same configurations in the model space over the entire sequence [4] or to subtract the explicit configuration-interaction from the effective parameter. Proceeding in this way, we conclude that it is possible to meaningfully combine perturbation theory (using effective parameters) and a diagonalization approach (using interaction integrals) into one orthogonal operator description of an atomic system. The convincing comparison of recent branching fractions and $log(gf)$ measurements to our calculations in Co II [25] illustrates this point.

Another case of strong CI occurs for higher ionization in the odd system, when the p-shell opens up. The resulting wide $3p^53d^{N+1}$ configurations seem to be a problem for any atomic physics approach [26], even for the impressive large scale MCDHF calculations that have recently been undertaken [27,28]. We are presently developing orthogonal operators suited for those systems, with special attention to the particularly large magnetic configuration interactions involving $\zeta(3p, np)$.

2.4. Transition Probabilities Improved

To obtain relatively accurate transition probabilities, both the eigenvectors to be used in the intermediate coupling transformation and the transition integrals should be optimized. While an orthogonal operator LSF is already optimized to produce satisfactory eigenvectors, work has to be done to obtain reliable transition integrals and to incorporate the most important core and valence excitations. To achieve this, several steps are taken:

- Use of core-polarization to account for the induced dipole moment, which is particularly important in the case of large, loosely bound (lower ionization stages) ionic cores.
  This usually decreases the E1-integral by 5–10%: $\vec{d} \rightarrow \vec{d}\left(1 - \frac{\alpha_d}{r^3}\right)$ where the dipole polarizability α_d (in terms of a_0^3) is either taken from literature or calculated *ab initio* . A cutoff radius is introduced

here to avoid divergence at $r = 0$.

For E2 transitions, the electric quadrupole polarization α_q is used.

- Use of MCDHF calculated transition integrals.
- Inclusion of essential configurations in the model space for full diagonalization.
- Use of perturbation theory: let Ψ and Ψ' refer to the full odd and even states of the system, to be approximated by the model states α and α' respectively, and $\beta, \gamma..$ far-lying configurations to be summed over. The first order expression $\langle \alpha' | \mathbf{r} | \alpha \rangle$ of the dipole operator $\mathbf{r}$ can be corrected to second order by linking the virtual configurations β, γ to the model configurations α, α' with the Coulomb operator V:

$$\langle \Psi' | \mathbf{r} | \Psi \rangle = \langle \alpha' | \mathbf{r} | \alpha \rangle + \sum_\beta \frac{\langle \alpha' | \mathbf{r} | \beta \rangle \langle \beta | V | \alpha \rangle}{E_\alpha - E_\beta} + \sum_\gamma \frac{\langle \alpha' | V | \gamma \rangle \langle \gamma | \mathbf{r} | \alpha \rangle}{E_{\alpha'} - E_\gamma} \tag{6}$$

To elaborate further on the last two points: with orthogonal operators, energy effects of unobserved configurations are accounted for by effective operators, while the number of strongly interacting configurations required in the model space is limited. Effects of unobserved configurations on the transition probabilities, however, are not automatically included in this way. This means that core-polarization effects are not absorbed by orthogonal operators and have to be included as corrections to the transition integrals. Also, second order corrections to the transition matrix elements may be added to take into account the valence effects of large numbers of far-away configurations [29]. The new radial factors entering this approach involve complete channels of single electron orbitals and can effectively be calculated with a B-splines program based on a frozen HF core [23,24].

Below, we use the $3d \rightarrow 4s$ excitation as a example: $\langle \alpha' | = \langle 3d^9 |, |\alpha\rangle = |3d^8 4p\rangle$ and $|\gamma\rangle = |3d^8 4s\rangle$:

$$\langle \Psi' | \mathbf{r} | \Psi \rangle \approx \langle 3d^9 | \mathbf{r} | 3d^8 4p \rangle + \frac{\langle 3d^9 | V | 3d^8 4s \rangle \langle 3d^8 4s | \mathbf{r} | 3d^8 4p \rangle}{E_{3d} - E_{4s}} \tag{7a}$$

The above expression turns out to give good agreement with a full diagonalization approach:

$$\langle \Psi' | \mathbf{r} | \Psi \rangle \approx \langle 3d^9 + 3d^8 4s | \mathbf{r} | 3d^8 4p \rangle \tag{7b}$$

Another important example of this point concerns the excitations $3s \rightarrow 3d$ and $3p^2 \rightarrow 3d^2$ within the Layzer complex $n = 3$: especially for higher ionization they should be included [30] either explicitly or by the above perturbative approach of equation (6).

3. Origin

To get a feel for the increased stability of orthogonal operators in a LSF and why they are least correlated, one may look at the simple high school project of Figure 2 for comparison.

To find the center of mass of an extended object like a bicycle, one may suspend this object under various angles, determine the plumb line in each case and preserve them like the yellow lines in the picture. The intersection of the plumb lines is the center of mass. Mathematically, this means finding the best intersection of a number of straight lines, a problem to be solved with least squares. Orthogonal plumb lines (being most independent) do the most accurate job.

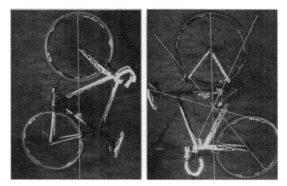

Figure 2. Determining the center of mass of a bicycle: orthogonal lines are the most accurate.

We recall a little linear algebra to describe the fitting process in a simplified way. Let P_i be one parameter of a vector $\mathbf{P}$ of parameters and E_k one energy of a vector $\mathbf{E}$ of energies. All energy operators will be contained in the matrix A.

To solve for $\mathbf{P}$, multiply with the transpose A^T from the left and invert:

$$A\mathbf{P} = \mathbf{E} \rightarrow \mathbf{P} = \left(A^T A\right)^{-1} A^T \mathbf{E} \tag{8}$$

The matrix $\left(A^T A\right)_{ij}$ is now the matrix of the inner products of all operators: $\left(A^T A\right)_{ij} = p_i : p_j$. For orthogonal operators, this matrix is diagonal and even orthonormal if $A^T = A^{-1}$. Thus, the final solution in the orthogonal case becomes:

$$\mathbf{P} = A^T \mathbf{E} \quad \text{or in index notation:} \quad P_i = \sum_k a_{ik} E_k \tag{9}$$

As there is no reference in the solution of P_i to any other parameter P_j, they turn out to be completely uncorrelated! Actually, this derivation only holds for diagonal matrix elements, which explains why orthogonal operators are least correlated instead of uncorrelated.

Orthogonal operators are normalized in batches of operators of the same type (2-body magnetic, 3-body electrostatic, etc.), which means that $\left(A^T A\right)_{ii} = p_i : p_i$ is the same for all operators in the batch. From the formula for the LSF error on the parameter P_i:

$$\delta P_i = \sqrt{\left(A^T A\right)_{ii}^{-1}} \cdot \sigma \tag{10}$$

it can now be understood, why parameter errors are equal (and minimal) for all parameters in the same batch.

3.1. Construction of an Orthogonal Set

First, let us make a subdivision of possible energy operators, to be able to survey the field:

1. There are three subspaces of operators that are orthogonal by their tensorial character: expressed as double tensors with ranks $k = 0, 1, 2$ in separate spin- and orbital spaces [31], one distinguishes: $T^{(00)0} \rightarrow$ electrostatic, $T^{(11)0} \rightarrow$ spin-orbit and $T^{(22)0} \rightarrow$ spin-spin.
2. Operators acting on different electrons belong in different orthogonal subspaces as well. The $\ell - \ell$ and $\ell - \ell'$ interactions are described, for example, by separate orthogonal operators.
3. In addition, each operator has a unique n–particle character, i.e., the number of electrons it acts on (only the average energy is a 0-particle operator). We distinguish $n = 2, 3, 4$ in the electrostatic space, $n = 1, 2, (3)$ in the spin-orbit space and $n = 2$ in the spin-spin space. An operator may

have different n–particle characters in different shells: the Trees operator T_1 has a three-particle character in the d-shell, while the T_{dds} operator has a two-particle character in the d-shell and a 1-particle character in the s-shell.

4. A further classification is the order of perturbation theory: preferably, we describe first- and second (or higher) order effects by different operators. In line with the previous point: n-body operators occur in the $(n-1)$ order of perturbation.

There are some useful properties of inner products that help defining a set of orthogonal operators. First, the inner product is independent of the coupling scheme.

Second, the behavior of the inner product as a function of the number of electrons in the shell is well-defined. If operators H_1 and H_2 occur together in the l^n shell for the first time, then their inner product in the l^N shell is closely related:

$$H_1 : H_2^{(N)} = \alpha \; \text{Tr} \; H_1^{(n)} \cdot \text{Tr} \; H_2^{(n)} + \beta \; H_1 : H_2^{(n)} \tag{11}$$

The coefficients α and β only depend on N and on the n–particle characters of the two operators: α and β are independent of the operators in question. In an orthogonal operator set, only the average energy operator has a non-zero trace. As a result, once operators are orthogonal in their parent configuration, i.e. the shell(s) where they first make their appearance, then they automatically remain orthogonal in all other configurations. This statement is equivalent to the below group theoretical result [32]:

If H_1 and H_2 belong to different irreducible representations Γ_1 and Γ_2 (differing symmetries) of a group $\mathcal{G}$ ánd $H_1 H_2$ does not contain the identity representation Γ_0 of $\mathcal{G}$, then $\rightarrow H_1 : H_2 = 0$.

This property has been used notably by Brian Judd to construct orthogonal operators based on Lie groups such as $U(4\ell + 2)$, $Sp(4\ell + 2)$, $SO(2\ell + 1)$ and G_2 [33,34].

Except building operators with well-defined group-theoretical properties, one may also start from elementary building blocks for inequivalent electrons that are orthogonal due to the well-known properties of $9j$-symbols. To build a first orthogonal basis, we use annihilation and creation tensor operators [31] with ranks κ and k in spin- and orbital spaces, coupled to a total rank t:

$$\left(\ell\ell'(SL) \,\|\, \left\{ \left(a^\dagger a \right)^{\kappa k} \left(b^\dagger b \right)^{\kappa' k'} \right\}^{tt} \,\|\, \ell\ell'(S'L') \right) = [t] \, [S, L, S', L']^{1/2} \, [\kappa, k, \kappa', k']^{1/2} \begin{Bmatrix} \frac{1}{2} & \frac{1}{2} & \kappa \\ \frac{1}{2} & \frac{1}{2} & \kappa' \\ S & S' & t \end{Bmatrix} \begin{Bmatrix} \ell & \ell & k \\ \ell' & \ell' & k' \\ L & L' & t \end{Bmatrix} \tag{12}$$

In the electrostatic $t = 0$ case, this simplifies to $\kappa = \kappa', k = k'$ yielding a number of $(4l' + 2)$ $e_{\kappa k}$ basic orthogonal operators in $ll'(l' < l)$ [35].

Such a first orthogonal basis of six electrostatic operators is given in Table 9 for the dp configuration as an example:

Table 9. Matrix elements of operators $e'_{\kappa k}$ for dp.

	e'_{00}	e'_{10}	e'_{01}	e'_{11}	e'_{02}	e'_{12}
^1P	1	3	3	9	7	21
^1D	1	3	1	3	−7	−21
^1F	1	3	−2	−6	2	6
^3P	1	−1	3	-3	7	-7
^3D	1	−1	1	−1	−7	7
^3F	1	−1	−2	2	2	−2
$\eta_{\kappa k}$	1	$\sqrt{3}$	2	$2\sqrt{3}$	$2\sqrt{7}$	$2\sqrt{21}$

In order to avoid square roots in the entries, the common normalization factors $\eta_{\kappa k}$ are given below each column: $e'_{\kappa k} = \eta_{\kappa k} \cdot e_{\kappa k}$.

The next step is to find linear combinations of these operators to distinguish between first order direct and exchange (F^k, G^k) Coulomb interactions and higher order effects. The Coulomb parameters are named C_i and the distinct higher order parameters S_i (Sack), respectively [36,37]. Operators are always written lower case to distinguish them from the corresponding parameters, e.g., the e_{av} operator is associated with the parameter E_{av}.

The below example describes the route from the original orthogonal basis given in Table 9 towards the final orthogonal operator set used for $d^n p$ configurations [37] and may serve as a blueprint for any ll' configuration.

Properties: $f^k \propto e'_{0k}$ and $(e'_{0k} - e'_{1k}) : (3 \cdot e'_{0k} + e'_{1k}) = 0$.

Immediate use: $e_{av} = e'_{00}$, $c_1 = e'_{02}$ and $s_1 = (3 \cdot e'_{01} + e'_{11})$. Property: $g^k \propto (e'_{0k} - e'_{1k})$ applies to all exchange operators.

Use: we combine the remaining operators: e'_{10}, e'_{12} and $(e'_{01} - e'_{11})$ to: $(7e'_{10} + e'_{12})$ and its orthogonal counterpart $(4e'_{10} - e'_{12})$. To check the whole procedure afterwards, we verify the inner products:

$$(e'_{01} - e'_{11}) : (3e'_{01} + e'_{11}) = 0 \text{ and } (7e'_{10} + e'_{12}) : (4e'_{10} - e'_{12}) = 0.$$

Final results:

First order Coulomb: $c_1 = e'_{02}$, $c_2 = (7e'_{10} + e'_{12})$ and $c_3 = 11(e'_{01} - e'_{11}) - (4e'_{10} - e'_{12})$

Higher order: $s_1 = (3e'_{01} + e'_{11})$, and $s_2 = \frac{3}{2}(e'_{01} - e'_{11}) + 2(4e'_{10} - e'_{12})$.

These final results are summarized in Table 10.

Table 10. Matrix elements of the c_i and s_i operators for dp.

	e_{av}	c_1	c_2	c_3	s_1	s_2
^1P	1	7	42	−57	9	−27
^1D	1	−7	0	−55	3	63
^1F	1	2	27	38	−6	18
^3P	1	7	−14	63	3	15
^3D	1	−7	0	33	1	−19
^3F	1	2	−9	−42	−2	−10
η_i	1	$2\sqrt{7}$	$\sqrt{231}$	$2\sqrt{517}$	$2\sqrt{3}$	$2\sqrt{141}$

All entries in a column are to be divided by the factors η_i to ensure common normalization: $e_{av} : e_{av} = c_i : c_i = s_i : s_i = 60$.

3.2. Completeness

An issue sometimes raised in connection with orthogonal operators is the comparatively large number of parameters M versus the number of observed energy levels N. Usually N is equal or larger than M, and ideally maybe even much larger. As each orthogonal operator describes an independent physical effect, however, it may quickly be seen that with a small number of parameters, many effects are inevitably omitted and one can not hope to obtain a physically reliable fit. On the other hand, for each $(n\ell)$–configuration with one electron outside closed shells, we have two operators e_{av} and ζ_l and therefore: $N = M = 2$: for this case we seem to be used to a complete set already!

In fact, each complete set of operators may be shown to yield a unique joint solution to level energies and level compositions. Consequently, an operator set that consists of more operators than the number of levels in the configuration is actually not overcomplete. In principle there is, in addition to the level energies, sufficient physical information dependent on the level compositions (Landé g-factors, line strengths) to determine all parameter values unambiguously. In many cases the experimental information is far from complete, but theoretical or empirical knowledge of the parameters can readily be used to reduce the number of parameters to be varied. For a Hamiltonian consisting of angular operators and associated radial parameters to yield correct energies and level compositions, a complete operator set should be used as a fact of principle, even though the number of operators may exceed the number of fitted energy levels. Noble-gas configurations p^5s, using Landé g-factors for additional information on the level compositions, have been used to substantiate this point [38]. The fit with $M = 5$ and $N = 4$ yielded physically realistic parameter values in line with *ab initio* results. In practice, however, one can always neglect the smaller effects, or add them as non-variable quantities derived from empirical or theoretical knowledge.

4. Outlook and Summary

The orthogonal operator approach is briefly reviewed and compared to the conventional LSF approach. The increased stability of orthogonal operators creates room to meaningfully include two-body magnetic operators and higher order effective electrostatic operators. The mean LSF error is thereby substantially reduced, which should give better eigenvector compositions and improved transition probabilities. However, 'orthogonal operators' is certainly no plug-and-play method: the initial estimates require iso-ionic/iso-electronic extrapolations, preliminary *ab initio* calculations or both. Experience with neighboring spectra obviously helps. Little experience has been gained in the open p^n and the f^n shells yet, though we recently implemented orthogonal operators for both cases; applications for f^n configurations with $n > 2$ are planned in the near future.

While large-scale calculations with Cowan's code (including many individual configurations) certainly lead in a quick and reliable way to satisfactory results, important magnetic effects are left out and this may constitute a problem for close lying levels. Orthogonal operators (including many effective operators) are more perturbative than variational in nature. As to higher order electrostatic effects, it is not easy to compare the impact of a large number of unobserved, scaled configurations (that indeed act as effective operators) to the effective 3- and 4-body orthogonal operators: they probably represent the same effects only partially. When strong configuration interaction comes into play, the orthogonal model space is extended with a limited number of configurations. The LSF mean error is still clearly smaller but closer to the mean error of the conventional approach in these cases.

We look forward to cooperate both with experimental groups to support their work and with theoretical groups to be able to implement more advanced *ab initio* methods. In the course of doing this, we hope to make the method of orthogonal operators more generally accessible.

Author Contributions: Both authors contributed to all aspects of this work. Writing–original draft, P.U. and T.R.; Writing–review and editing, P.U. and T.R.

Funding: This research received no external funding.

Conflicts of Interest: The authors declare no conflict of interest.

References

1. Judd, B.; Hansen, J.; Raassen, A. Parametric fits in the atomic d shell. *J. Phys. B At. Mol. Phys.* **1982**, *15*, 1457–1472. [CrossRef]
2. van het Hof, G.; Raassen, A.; Uylings, P. Parametric Description of $3d^N4s$ Configurations using Orthogonal Operators. *Phys. Scr.* **1991**, *44*, 343–350.
3. Hartmann, H.; Nilsson, H.; Engström, L.; Lundberg, H. The FERRUM project: Experimental lifetimes and transition probabilities from highly excited even 4d levels in Fe II. *Astron. Astrophys.* **2015**, *584*, 1–6. [CrossRef]
4. Azarov, V. Parametric study of the $5d^3$, $5d^2$ 6s and $5d^2$ 6p configurations in the Lu I isoelectronic sequence (Ta III-Hg X) using orthogonal operators. *At. Data Nucl. Data Tables* **2018**, *119*, 193–217. [CrossRef]
5. Raassen, A.; Uylings, P. 2019. Available online: https://personal.sron.nl/~tonr/atomphys/levtext.html (accessed on 11 November 2019).
6. Cowan, R. *The Theory of Atomic Structure and Spectra*; University of California Press: Berkeley, CA, USA, 1981.
7. Kramida, A. Cowan Code: 50 Years of Growing Impact on Atomic Atomic Physics. *Atoms* **2019**, *7*, 64. [CrossRef]

8.   Hansen, J.; Uylings, P.; Raassen, A. Parametric Fitting with Orthogonal Operators. *Phys. Scr.* **1988**, *37*, 664–672. [CrossRef]

9.   Hautecoeur, L. *De Impressionisten, Georges Seurat*; Atrium: Alphen aan den Rijn, The Netherlands, 1972.

10. Tchang-Brillet, W.Ü.; Artru, M.C.; Wyart, J.F. The $3d^4$-$3d^34p$ Transitions of Triply Ionized Manganese (Mn IV). *Phys. Scr.* **1986**, *33*, 390–400. [CrossRef]

11. Azarov, V.; Gayasov, R. The third spectrum of rhenium (Re III): Analysis of the $(5d^5 + 5d^46s)$–$(5d^46p + 5d^36s6p)$ transition array. *At. Data Nucl. Data Tables* **2018**, *121*, 306–344. [CrossRef]

12. Azarov, V.; Tchang-Brillet, W.Ü.; Gayasov, R. Analysis of the spectrum of the $(5d^6 + 5d^56s)$ - $(5d^56p+5d^46s6p)$ transitions of two times ionized osmium (Os III). *At. Data Nucl. Data Tables* **2018**, *121*, 345–377. [CrossRef]

13. Raassen, A.; Uylings, P. The Use of Complete Sets of Orthogonal Operators in Spectroscopic Studies. *Phys. Scr.* **1996**, *T65*, 84–87. [CrossRef]

14. Ekberg, J. Wavelengths and transition probabilities of the $3d^6$-$3d^54p$ and $3d^54s$-$3d^54p$ transition arrays of Fe III. *Astron. Astrophys. Suppl. Ser.* **1993**, *101*, 1–36.

15. Corrégé, G.; Hibbert, A. The Oscillator Strengths of Fe II $\lambda\lambda2507, 2509$. *Astrophys. J.* **2005**, *627*, L157–L159. [CrossRef]

16. Hibbert, A. Successes and Difficulties in Calculating Atomic Oscillator Strengths and Transition Rates. *Galaxies* **2018**, *6*, 77. [CrossRef]

17. Raassen, A.; Uylings, P. Critical evaluation of calculated and experimental transition probabilities and lifetimes for singly ionized iron group elements. *J. Phys. B At. Mol. Opt. Phys.* **1998**, *31*, 3137–3146. [CrossRef]

18. Kurucz, R. 2017. Available online: http://kurucz.harvard.edu/atoms/2601/gf2601.pos (accessed on 1 June 2019).

19. Verner, E.; Gull, T.; Bruhweiler, F.; Johansson, S.; Ishibashi, K.; Davidson, K. THE origin of Fe II and [Fe II] emission lines in the 4000–10000 Å range in the BD weigelt blobs of η CARINAE. *Astrophys. J.* **2002**, *581*, 1154–1167. [CrossRef]

20. Uylings, P. Extended theory of the spin-orbit interaction. *J. Phys. B At. Mol. Opt. Phys.* **1989**, *22*, 2947–2961. [CrossRef]

21. Uylings, P. Applications of second quantization in the coupled form. *J. Phys. B At. Mol. Opt. Phys.* **1992**, *25*, 4391–4407. [CrossRef]

22. Parpia, F.; Froese Fischer, C.; Grant, I. GRASP92: A package for large-scale relativistic atomic structure calculations. *Comput. Phys. Comm.* **1996**, *94*, 249–271. [CrossRef]

23. Hansen, J.; Bentley, M.; van der Hart, H.; Landtman, M.; Lister, G.; Shen, Y.T.; Vaeck, N. The Introduction of B-Spline Basis Sets in Atomic Structure Calculations. *Phys. Scr.* **1993**, *T47*, 7–17. [CrossRef]

24. Bachau, H.; Corbier, E.; Decleva, P.; Hansen, J.; Martín, F. Applications of B-splines in atomic and molecular physics. *Rep. Prog. Phys.* **2001**, *64*, 1815–1942. [CrossRef]

25. Lawler, J.; Feigenson, T.; Sneden, C.; Cowan, J.; Nave, G. Transition Probabilities of Co II Weak Lines to the Ground and Low Metastable Levels. *Astron. Astrophys. Suppl. Ser.* **2018**, *238*, 1–17. [CrossRef] [PubMed]

26. Ryabtsev, A. Survey of Some Recent Experimental Analysis of $3p^53d^{N+1}$ Configurations and of Rh I-like Spectra. *Phys. Scr.* **1996**, *1996*, 23–30. [CrossRef]

27. Li, Y.; Xu, X.; Li, B.; Jönsson, P.; Chen, X. Multiconfiguration Dirac–Hartree–Fock calculations of energy levels and radiative rates of Fe VII. *Mon. Not. R. Astron. Soc.* **2018**, *479*, 1260–1266.

28. Li, B.; Xu, X.; Chen, X. Relativistic large scale CI calculations of energies, transition rates and lifetimes in Ca-like ions between Co VIII and Zn XI. *At. Data Nucl. Data Tables* **2019**, *127*, 131– 139. [CrossRef]

29. Uylings, P.; Raassen, A. Accurate calculation of transition probabilities using orthogonal operators. *J. Phys. B At. Mol. Opt. Phys.* **1995**, *28*, L209–L212. [CrossRef]

30. Quinet, P.; Hansen, J. The influence of core excitations on energies and oscillator strengths of iron group elements. *J. Phys. B At. Mol. Opt. Phys.* **1995**, *28*, L213 – L220. [CrossRef]

31. Judd, B. *Second Quantization and Atomic Spectroscopy*; John Hopkins Press: Baltimore, MD, USA, 1967.

32. Judd, B. *Operator Averages and Orthogonalities*; Springer: Berlin, Germany, 1984; Volume 201.

33. Judd, B.; Leavitt, R. Many-electron orthogonal scalar operators in atomic shell theory. *J. Phys. B At. Mol. Phys.* **1986**, *19*, 485–499. [CrossRef]

34. Dothe, H.; Judd, B. Orthogonal operators applied to term analysis residues for Fe VI $3d^24p$. *J. Phys. B At. Mol. Phys.* **1987**, *20*, 1143–1151. [CrossRef]

35. Dothe, H.; Judd, B.; Hansen, J.; Lister, G. Orthogonal scalar operators for $p^N d$ and pd^N. *J. Phys. B At. Mol. Phys.* **1985**, *18*, 1061–1080. [CrossRef]
36. Klinkenberg, P.; Uylings, P. The $5f^2$-Configuration in Doubly Ionized Thorium, Th III. *Phys. Scr.* **1986**, *34*, 413–422. [CrossRef]
37. Uylings, P.; Raassen, A. High Precision Calculation of Odd Iron-group Systems with Orthogonal Operators. *Phys. Scr.* **1996**, *54*, 505–513. [CrossRef]
38. van het Hof, G.; Uylings, P.; Raassen, A. On the necessity and meaning of complete sets of orthogonal operators in atomic spectroscopy. *J. Phys. B At. Mol. Opt. Phys.* **1991**, *24*, 1161–1173.

 *atoms*

MDPI

Article

The Belgian Repository of Fundamental Atomic Data and Stellar Spectra (BRASS)

Alex Lobel [1,*], **Pierre Royer** [2], **Christophe Martayan** [3], **Michael Laverick** [2], **Thibault Merle** [4], **Mathieu Van der Swaelmen** [4], **Peter A. M. van Hoof** [1], **Marc David** [5] and **Herman Hensberge** [1] and **Emmanuel Thienpont** [6]

[1] Royal Observatory of Belgium, Ringlaan 3, B-1180 Brussels, Belgium; p.vanhoof@oma.be (P.A.M.v.H.); hhensberge@gmail.com (H.H.)
[2] Instituut voor Sterrenkunde, KU Leuven, Celestijnenlaan 200D, Box 2401, 3001 Leuven, Belgium; pierre.royer@ster.kuleuven.be (P.R.); mike.laverick@kuleuven.be (M.L.)
[3] European Organisation for Astronomical Research in the Southern Hemisphere, Alonso de Córdova 3107, Vitacura, 19001 Casilla, Santiago de Chile, Chile; cmartaya@eso.org
[4] Institut d'Astronomie et d'Astrophysique, Université Libre de Bruxelles, Av. Franklin Roosevelt 50, CP 226, 1050 Brussels, Belgium; tmerle@ulb.ac.be (T.M.); mathieu@arcetri.inaf.it (M.V.d.S.)
[5] Onderzoeksgroep Toegepaste Wiskunde, Universiteit Antwerpen, Middelheimlaan 1, 2020 Antwerp, Belgium; marc.david@uantwerpen.be
[6] Vereniging voor Sterrenkunde, Kapellebaan 56, 2811 Leest, Belgium; emmanuel.thienpont@gmail.com
* Correspondence: Alex.Lobel@oma.be

Received: 15 October 2019; Accepted: 19 November 2019; Published: 22 November 2019

Abstract: Background: BRASS (Belgian Repository of Fundamental Atomic Data and Stellar Spectra) is an international networking project for the development of a new public database providing accurate fundamental atomic data of vital importance for stellar spectroscopic research. We present an overview of research results obtained in the past four years. Methods: The BRASS database offers atomic line data we thoroughly tested by comparing theoretical and observed stellar spectra. We perform extensive quality assessments of selected atomic input data using advanced radiative transfer spectrum synthesis calculations, which we compare to high-resolution Mercator-HERMES and ESO-VLT-UVES spectra of F-, G-, and K-type benchmark stars observed with very high signal-to-noise ratios. We have retrieved about half a million atomic lines required for our detailed spectrum synthesis calculations from the literature and online databases such as VAMDC, NIST, VALD, CHIANTI, Spectr-W^3, TIPbase, TOPbase, SpectroWeb. Results: The atomic datasets have been cross-matched based on line electronic configuration information and organized in a new online repository called BRASS. The validated atomic data, combined with the observed and theoretical spectra are also interactively offered in BRASS. The combination of these datasets is a novel approach for its development providing a universal reference for advanced stellar spectroscopic research. Conclusion: We present an overview of the BRASS Data Interface developments allowing online user interaction for the combined spectrum and atomic data display, line identification, atomic data accuracy assessments including line $\log(gf)$-values, and line equivalent width measurements.

Keywords: quantitative stellar spectroscopy; spectral lines; atomic line data; atomic and spectral databases

1. Introduction

Fundamental atomic transition data, such as line oscillator strength values, are of central importance for determining the physical conditions in stellar atmospheres and for measuring their chemical compositions. Despite the significant work underway to produce these atomic data values for many astrophysically important ions, the uncertainties in these parameters remain large and can propagate throughout the entire field of astronomy. The Belgian repository of fundamental atomic data and stellar spectra (BRASS) aims to provide a large systematic and homogeneous quality assessment of the atomic data available for quantitative stellar spectroscopy. BRASS compares theoretical spectrum calculations to very high-quality observed spectra of FGK-type stars in order to critically evaluate the atomic data available for over a thousand atomic lines.

We report on the detailed analysis of six BRASS FGK-type benchmark HERMES spectra and the solar FTS spectrum. We present results of our quality assessments of atomic line oscillator strengths ($\log(gf)$-values) and line rest-wavelengths we have collected and combined in BRASS for advanced theoretical spectrum calculations of the benchmark spectra. Section 2.1 discusses the cross-matching of atomic transitions retrieved from a variety of atomic databases and the literature for the development of the Lines BRASS Data Interface. In Sections 2.2 and 2.3 we discuss the benchmark spectrum modeling results of 1091 investigated atomic lines for the $\log(gf)$ accuracy assessment pages we offer in the Spectra BRASS Data Interface. Section 2.4 provides a results comparison of two atomic data quality assessment methods used in BRASS. Section 2.5 presents a concise multiplet analysis of investigated Fe I transitions. The summary is provided in Section 3.

2. BRASS Development Status

2.1. Lines BRASS Data Interface

An important source of uncertainty in stellar spectrum synthesis calculations is the accuracy of atomic data of permitted transitions. It is crucial to constrain atomic data uncertainties for reliable measurements of the thermal conditions and chemical composition of stellar atmospheres. For BRASS we retrieve $\sim$400,000 transition entries from various online atomic databases: VALD-3, NIST, Spectr-W^3, TIPbase, TOPbase, CHIANTI, and SpectroWeb. We collect the atomic transition data of neutral species and ions up to 5$^+$ for wavelengths between 420 and 680 nm. The datasets are homogenized and cross-matched against the BRASS atomic line list compilation. The BRASS list is composed of Kurucz and NIST V4.0 lines containing for each transition the species (element and ionization stage), line rest-wavelength, $\log(gf)$, upper and lower electronic configurations and energy levels, J-values, and the corresponding literature references. Our cross-matching is performed in two different ways: the parametric cross-match method is based on wavelength- and level energy-values for finding the same transition of a given species. On the other hand, the non-parametric cross-match method is based on detailed electronic configuration information for finding transitions that are physically identical between the datasets. The cross-matching accounts for atomic fine structure, the provided isotopic information, and the type of transition. It however does not account for currently missing hyperfine structure information.

The BRASS compilation was initially tested with theoretical spectrum calculations of the solar flux spectrum [1] and using Mercator-HERMES [2] spectra of selected B-, A-, F-, G-, and K-type stars (see [3]). The BRASS list has been also cleaned from numerous un-observed lines, spurious atomic & molecular background features, and duplicated lines have been excluded. Note that the SpectroWeb atomic lines list was previously compiled from VALD-2 and NIST data (V2.0 through V4.0), and was also extensively tested similar to the BRASS list with theoretical spectrum calculations of high-quality hot and cool star spectra [4]. Table 1 lists the number of retrieved lines, source databases, dates of retrieval, and various atomic data values collected from each database. We have made extensive use of the online VAMDC portal offering homogenized datasets which has expedited the comparison and cross-matching of the datasets we have retrieved for BRASS. We partly incorporate data from TIPbase and TOPbase and include some of our expansions into fine-structure transitions [5]. We also calculate line $\log(gf)$-values for Spectr-W^3 using the f_{ik}-values they offer.

Table 1. Overview of the retrieved number of lines for BRASS from various atomic databases, including the retrieval dates and types of atomic data per source.

Data Source	Origin	No. Lines	Date	Species	Line Wavelength	A_{ki}	f_{ik}	$\log(gf)$	$E_{low/up}$	$J_{low/up}$
BRASS	-	82,337	2009–2012	✓	✓			✓	✓	✓
SpectroWeb	-	62,181	2004–2008	✓	✓			✓	✓	
VALD-3	VALD	158,861	May 2016	✓	✓			✓	✓	✓
NIST	NIST	36,123	Mar 2016	✓	✓	✓	✓	✓	✓	✓
Spectr-W^3	VAMDC	5515	Mar 2016	✓	✓	✓	✓		✓	✓
TIPbase	NORAD	33,108	Feb 2017	✓	✓	✓	✓		✓	
TOPbase	VAMDC	33,462	May 2016	✓	✓			✓	✓	
CHIANTI	VAMDC	3587	Mar 2016	✓	✓	✓		✓	✓	✓

For the BRASS project [6] have used the non-parametric cross-match method to explore differences between multiple occurrences of identical transitions in the retrieved datasets. Detailed comparisons of λ vs. $\Delta\lambda$, E vs. ΔE, $\Delta\lambda$ vs. ΔE, and $\Delta\lambda$ vs. $\Delta\log(gf)$-values mainly reveal the presence of small-scale conversion precision differences. Large-scale systematic correlations are detected for a few cases only. However, the comparison of the line $\log(gf)$-values reveals differences in excess of 2 dex, which has important implications for quantitative stellar spectroscopy.

An investigation of duplicated transitions (also accounting for hyperfine-, isotopic-, and E2-M1 forbidden-transitions) in the retrieved datasets show a significant number of almost 2% in VALD-3 lists. These duplicates could be sourced back to the original work in 99% of cases, hence they were not produced by the databases from which the BRASS datasets are retrieved. The duplicated transitions for example have not been detected in the line datasets retrieved from NIST.

The cross-matched atomic datasets, including the BRASS atomic lines compilation, have been incorporated in the online Lines BRASS Data Interface (LBDI) at brass.sdf.org. Lists of duplicated lines are also offered there for a variety of data formats (HTML, ASCII, PDF). The left-hand panel of Figure 1 shows the LBDI that can be queried for a given element in a user-defined wavelength interval. In case a cross-match listing for every element is requested the users can set the Element input field to *all*. The query results can be sorted by increasing rest-wavelengths or $\sigma\log(gf)$ (standard deviation)-values marked in blue in Figure 2. The query results can be exported and saved in extensive line lists or per user-selected line to machine-readable (tab-separated ASCII) tables. Figure 2 shows for example the cross-matched atomic data of a S II line retrieved from seven atomic data sources providing five different $\log(gf)$-values ranging from -0.341 dex to -0.059 dex. The literature references of the $\log(gf)$-values are offered together with the upper and lower electronic configurations and level energies. The right-hand panel of Figure 1 also shows a subset of LBDI dynamic plots of the BRASS compilation $\log(gf)$-values vs. $\log(gf)$-difference values for VALD3-BRASS and NIST-BRASS of cross-matched Fe I lines and of Fe II lines. These dynamic plots can be interactively zoomed and the data of individual lines marked and displayed by mouse interaction. The $\log(gf)$-difference plots are provided per query for all atomic data sources and are ordered by neutral, singly , and multiply ionized species (from left to right). This provides users with an interactive and comprehensive overview of all cross-matched $\log(gf)$ datasets offered in BRASS. Note that the BRASS Data Interface also offers comprehensive Help pages (under the main green tab) for a number of BRASS usecases and corresponding tutorial videos.

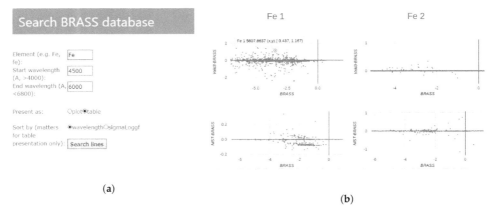

Figure 1. (**a**) The left-hand panel shows the Lines BRASS Data Interface query page. (**b**) The plot panels to the right show LBDI dynamic plots of BRASS log(gf)-values compared to the log(gf)-differences for VALD3-BRASS (top plot panels) and NIST-BRASS (bottom plot panels) of cross-matched Fe I and of Fe II lines (*see text*).

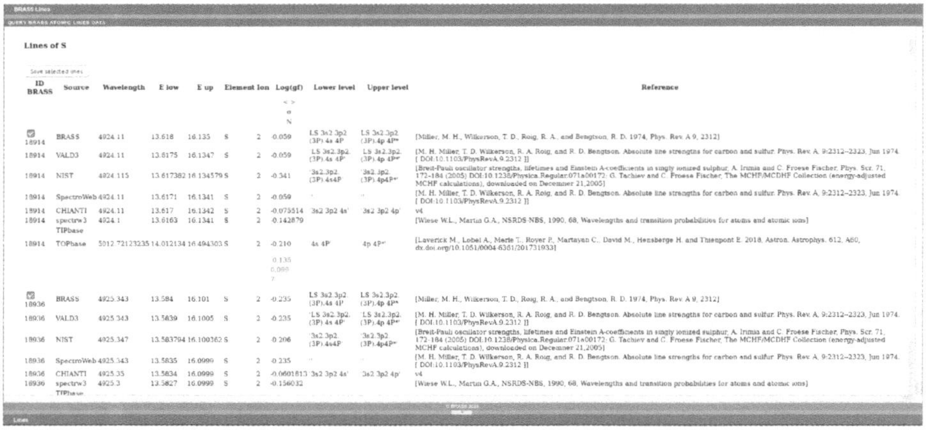

Figure 2. BRASS query results table for S lines sorted by wavelength. The lines data are cross-matched with various atomic databases and the corresponding literature references are also offered.

2.2. Spectra BRASS Data Interface

For the BRASS project we observe benchmark spectra of a variety of bright stars ($V < 7^m$) with the HERMES and ESO-VLT-UVES high-resolution spectrographs. We investigate HERMES benchmark spectra of 6 dwarf stars of F-, G-, and K spectral types observed with very high signal-to-noise ratios (SNR) of ~800–1000: 51 Peg, 70 Oph, 70 Vir, 10 Tau, ϵ Eri, and β Com (T_{eff} between 5000 K and 6000 K). The spectra are modelled in detail with advanced LTE synthesis calculations using 1-D hydrostatic atmosphere models (see [3]). The detailed spectrum modeling determines T_{eff}-, log(g)-, [M/H]-, ζ_μ-, $v\sin i$-, and [α/Fe]-values we also compare to published stellar parameters measured with high-resolution spectra. The BRASS benchmark stars exclude binaries and are selected for non-variability and non-peculiarity. They are normal dwarf stars with narrow absorption lines having small rotational velocities below 6 km s^{-1} and metallicities very close to solar values. Metal-poor stars are excluded to avoid non-LTE effects in our theoretical spectrum calculations.

The spectra of 11 BRASS benchmark stars (including the solar FTS spectrum) and the theoretical spectra are incorporated in the Spectra BRASS Data Interface (SBDI) shown in Figure 3. Users can interactively display up to four spectral Regions in two benchmark stars selected from the left-hand menus. The wavelengths of identified atomic lines are marked with red and blue labels. The red label numbers mark investigated lines. The red (and blue) labels can be clicked for displaying BRASS atomic data (log(gf)- and E_{low}-values in a red 'graded list' in the right-hand panels), together with measured line properties such as the observed line equivalent widths (W_λ and W_λ-error), and the type of quality assessment we performed for the accuracy of the line log(gf)- and rest-wavelength -values (see Section 2.3). By double-clicking the red line labels users can build up lists of BRASS line data values (marked in green) for saving to their local computer disc. Clicking the 'View data quality' link in the red (or green) tables populates the 'Atomic Data Quality' tab in the central SBDI panel for a complete overview of the atomic line data quality assessment results BRASS offers for the investigated line.

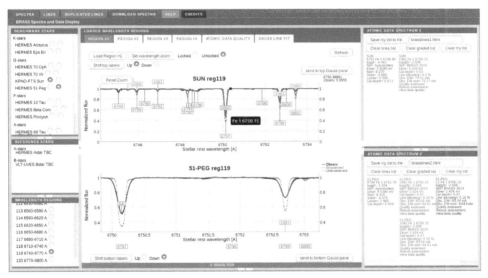

Figure 3. Screencopy of the SBDI for the solar FTS spectrum (central top panel) and the BRASS benchmark HERMES spectrum of the solar-like star 51 Peg (central bottom panel). The red and blue interactive labels mark identified absorption lines with atomic and line property data tables shown in the right-hand sub-panels. Atomic quality assessment pages are displayed under the central Atomic Data Quality tab by clicking on the View data quality link in the red or green line data tables.

2.3. Atomic Line Data Quality Assessments

A large-scale homogeneous selection of atomic lines was performed by [7] for BRASS by calculating the theoretical spectra of the 6 FGK benchmark HERMES spectra and the solar FTS spectrum. A selection of 1091 theoretically deep and sufficiently un-blended lines in the wavelength range 420 nm to 680 nm proved to be suitable for advanced quality assessments of the accuracy of the atomic datasets collected in BRASS. We determine astrophysical (semi-empiric) log(gf)-values for these 1091 transitions using two commonly employed analysis methods. The agreement of the measured log(gf)-values is used for selecting well-behaving lines for the quality assessment work. A total of 845 atomic lines are found to be suitable for quality assessment, of which 408 are robust against any systematic differences between both analysis methods. Around 53% of the quality-assessed lines are found to have at least one literature log(gf)-value in agreement with the calculated values, although the remaining values can disagree by as much as 0.5 dex (see Section 2.4).

For selecting atomic lines we calculate the amounts of blending in the 82337 BRASS lines of the solar and 51 Peg benchmark spectra. To reduce the impact of the line blending amounts on the atomic data quality assessment work a cut-off for blending of $\leq$10% is used, selected as a good balance between blending of the line core and the number of investigated lines. An additional cut-off on the central line core depth $\geq$0.02 is also used to ensure the observed line profiles can be measured with sufficient accuracy. A total of 1515 atomic lines is initially selected as 'un-blended' lines in both stars. The lines selection procedure does not place limits on the atomic species. The equivalent line widths of the 1515 un-blended lines are automatically measured in all seven benchmark spectra using a single Gaussian fit profile. The line fit procedure is optimized using Gauss-Newton non-linear regression, or Nelder-Mead minimization in the case of slow convergence [8]. The best fit to the observed line fluxes is limited to the wavelength interval between two local flux maxima in both line wings exceeding 2% of the normalized continuum flux level. Beyond the local flux maxima the W_λ integration is extended for Gaussian line wings. A goodness-of-fit value of $\chi^2 \leq 0.95$ is used to remove poorly fitted (in addition to visual inspection), non-existent, too blended, or Earth line contaminated absorption features. The SBDI also offers an interactive W_λ measurement tool under the Gauss line fit tab of the central panel in Figure 3. Users can select lines in observed BRASS spectra and display the single Gaussian best fit result together with the list of measured line properties for saving to their computer's disc.

The astrophysical $\log(gf)$-values are determined with two commonly employed methods. The measured line equivalent widths are converted into $\log(gf)$-values using the theoretical curve-of-growth calculated for the line in each benchmark star (called COG method). The other method varies the $\log(gf)$-values in detailed radiative transfer calculations for determining the best-fit value to the observed line profile. The latter method is called GRID because it involves an iterative line modeling procedure for which a grid of spectra is calculated and the best fitting spectrum for a range of $\log(gf)$- and λ-values is obtained with χ^2-minimization by interpolating in steps of $\Delta\lambda$ (0.005 Å) and $\Delta\log(gf)$ (0.01 dex) using a bivariate cubic spline fit. Both methods introduce assessable uncertainties resulting from the accuracy of the best fit procedures to the observed W_λ-value and the continuum normalized line flux distribution. The uncertainties can be attributed to the spectral SNR, specific atmosphere modeling assumptions, the continuum flux level normalization procedure, and blending with the observed line unaccounted for in our theoretical spectrum calculations. For the seven BRASS benchmark spectra we measure an intrinsic scatter between both methods of $\pm$0.04 dex (1σ standard deviation) for line blending levels below 3–4%. The value of $\pm$0.04 dex is therefore used as a constraint on the lines selection for limiting the impact of systematic differences between both methods on the atomic data quality assessment results. In our analysis method the close agreement between the COG and GRID $\log(gf)$-values is required for quality assessing the literature $\log(gf)$-values retrieved for BRASS. The COG $\log(gf)$-value is calculated for a given transition with the observed W_λ-value of an absorption feature we can attribute to the line, while the GRID $\log(gf)$-value results from complete theoretical spectrum calculations that fit the observed spectrum incorporating the (sufficiently un-blended) line profile.

The SBDI offers atomic data quality assessment pages showing plots and data values for the 1091 investigated lines. Figure 4 shows a screencopy of the SBDI Atomic Data Quality tab for the Ni I λ6598 line observed in the BRASS benchmark spectra (solid black line with dots) over-plotted with the theoretical profiles we calculate for the atomic data values retrieved from four atomic databases. The line profiles calculated for $\log(gf)$-values we determine from the GRID and COG analysis methods are over-plotted in blue and green colors, respectively. Users can interactively zoom-in, pan, and reset these line profile plots for each benchmark star. By clicking the check boxes above the plots the theoretical line profiles calculated with the $\log(gf)$-values in the atomic databases are also over-plotted for user inspection. The Quality assessment table shown below the line profile plots lists the GRID and COG line $\log(gf)$- and rest-wavelength (λ)-values, together with the differences ($\Delta\log(gf)$ and $\Delta\lambda$) with respect to the GRID values. The last column of this table offers a Yes/No flag indicating if the $\Delta\log(gf)$-value is within the errors of the GRID $\log(gf)$-value. The flags and Δ-values are useful for determining if the $\log(gf)$-values retrieved from the databases for BRASS are sufficiently accurate for detailed spectrum synthesis calculations. For example, for Ni I $\lambda6598$ we determine GRID $\log(gf)$- and COG $\log(gf)$-values within errors of each other (hence having quality-assessable atomic data), but not within ±0.04 dex of each other, signaling the line is not robust against the analysis method. The bottom table with Equivalent widths offers the observed (Measured) and theoretical W_λ-values (in mÅ) we calculate for the investigated line per database in all the benchmark stars. Note that we also add small corrections listed for ΔW_λ^{corr} to the observed line equivalent width values in case the line saturates on the curve-of-growth and Voigt profile corrections are introduced in our best Gaussian line fit procedure.

Figure 4. Screencopy of the SBDI page for the atomic data quality assessment results of the Ni I $\lambda6598$ line observed in seven BRASS benchmark spectra. The observed and theoretical line profiles shown in the sub-panels can be displayed with user interaction. The SBDI pages offer an overview of all atomic data quality results for each investigated line together with the observed and theoretical line equivalent width values (see text).

2.4. Comparison of Atomic Data Quality Assessment Results

We find 845 of the 1091 investigated lines to be quality-assessable, and 408 are also analysis-independent lines. Nearly half of the investigated and quality-assessable lines are of Fe I, while another ~10% belong to singly ionized species. The retrieved literature $\log(gf)$-values of a quality-assessable line are considered in agreement with our results and can be recommended in theoretical spectrum calculations only in case they agree within the errors of the mean (averaged over all benchmarks) GRID $\log(gf)$-value and its standard deviation. We do not consider any literature errorbars because they are not available for the vast majority of investigated lines. In most cases we adopt the mean GRID $\log(gf)$-value as the BRASS reference value because the GRID method yields smaller χ^2-values than the COG method. About 53% of the quality-assessable lines have literature $\log(gf)$-values in agreement with the mean GRID $\log(gf)$-values. A similar percentage of the 408 analysis-independent lines have sufficiently accurate atomic data. The majority of Fe-group species (V I, Cr I, Mn I, Co I, Ni I, Ti I, and Sc II, Ti II, Fe II) have a good number of lines with accurate atomic data for 70–75% of the lines. The Fe I lines, however, have only ~38% with sufficiently accurate atomic data (see Section 2.5).

The right-hand panel of Figure 5 shows mean GRID $\log(gf)$- (blue dots) and mean COG $\log(gf)$-values (black dots) compared to the $\log(gf)$-values in the BRASS (input) dataset for the

408 analysis-independent lines (where both COG and GRID astrophysical values agree within ±0.04 dex). We find sizable differences with the BRASS $\log(gf)$-values for a considerable number of lines. Difference $\log(gf)$-values in excess of ±0.5 dex are observed. The inset panel shows lines with smaller $\log(gf)$ differences (≤0.2 dex), although many are not in agreement within the derived errorbars.

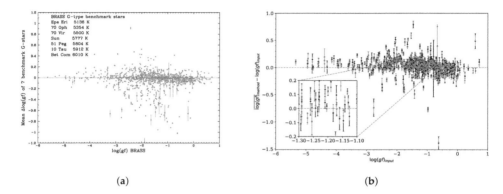

(a)   (b)

Figure 5. (a) The left-hand panel shows mean $\Delta\log(gf)$-values (red dots) determined with the linear approximation method against the literature $\log(gf)$-values retrieved for BRASS. The standard deviations (blue lines) for the 7 benchmark stars stay below 0.02 dex for the vast majority of investigated lines. (b) The right-hand panel shows a similar difference plot for astrophysical $\log(gf)$-values of 408 analysis-independent lines from the GRID (blue dots) vs. COG (black dots) atomic data quality assessment method (see text).

It is important to point out that the large differences between the literature $\log(gf)$-values we retrieve for BRASS and the mean astrophysical $\log(gf)$-values calculated with FGK BRASS benchmark spectra are also detected using a linear approximation method. Absorption lines on the linear part of the curve-of-growth follow a linear relationship between W_λ- and $\log(gf)$-values. For these lines the difference between observed and theoretical $\log(gf)$-values equals $\log(W_\lambda^{obs}/W_\lambda^{mod})$, where W_λ^{mod} is the line equivalent width we calculate with the theoretical $\log(gf)$-value.

The left-hand panel of Figure 5 shows the mean of the $\Delta\log(gf)$-values we calculate for the seven BRASS benchmark stars against the retrieved BRASS $\log(gf)$-values. The largest mean $\Delta\log(gf)$-values can also exceed 0.5 dex, although the standard deviations are ≤0.02 dex for the majority of investigated lines (blue errorbars) (see [8]). Note however that the mean COG and GRID errors are 0.065 dex and 0.05 dex, respectively, or about 3 times larger. Similar to the GRID vs. COG quality assessment method the mean $\log(gf)$-differences we calculate with this linear approximation method remain typically below ±1 dex and are chiefly observed for the medium-strong lines having $-3 \leq \log(gf) \leq -0.5$. The lines with negative $\log(gf)$-differences were also found in a separate analysis of Fe-group element lines in the solar FTS spectrum and in HERMES and UVES spectra of Procyon and ϵ Eri [4]. For these lines the literature $\log(gf)$-values are overestimated yielding theoretical W_λ-values that exceed observed values. Similar to the full-fledged GRID vs. COG analysis method smaller $\Delta\log(gf)$-values are also found towards the weakest ($\log(gf)<-3.5$) and strongest ($\log(gf)>0$) investigated lines.

2.5. Multiplet Analysis of Fe I Transitions in BRASS

The rather small percentage of only ∼38% of sufficiently accurate atomic data for the Fe I lines in BRASS calls for an investigation of its origin we briefly discuss. The left-hand panel of Figure 6 shows the curve-of-growth for Fe I lines we observe in the solar benchmark spectrum. The black dots show observed (reduced) W_λ/λ-values against $\log(gf)$ (co-added with other terms), for $\log(gf)$-values in the (input) BRASS compilation of Table 1. We find considerable scatter for the transitions on the linear part of the curve or mainly for the weak and medium-strong Fe I lines. The large scatter is due to the limited accuracy of the literature $\log(gf)$-values for these lines. We find that this scatter across the curve substantially reduces after replacing the literature $\log(gf)$-values with the ones we calculate from the linear approximation method in Section 2.4 shown with red dots. By replacing the $\log(gf)$-values with the ones we calculate from the COG vs. GRID method the scatter nearly vanishes and the curve assumes the smooth (and narrow) shape required for atomic lines belonging to the same species in stellar spectra. The large percentage we find of over 60% of literature Fe I atomic data with limited quality mainly results from medium-strong (and weak) lines having $-3 \le \log(gf) \le -0.5$ in Figure 5.

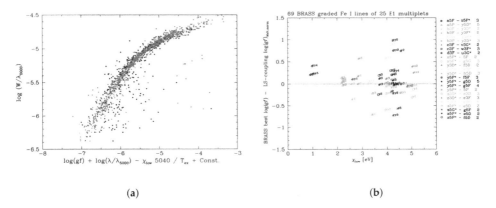

(a) (b)

Figure 6. (**a**) The left-hand panel shows the curve-of-growth of Fe I lines investigated for BRASS. Substantial scatter observed in medium-strong and weak lines using literature $\log(gf)$-values (black dots) is properly removed by replacing them with the values of the linear approximation method (red dots) or the GRID $\log(gf)$-values (blue dots). (**b**) The right-hand panel shows the increase in $\Delta\log(gf)$-values (BRASS-LS) against E_{low} for lines of 25 multiplets (also marked to the right for the number of lines) having literature retrieved $\log(gf)$-values of lesser quality in case $E_{low} > 4$ eV.

A more extensive analysis of the Fe I fine structure data we retrieve for BRASS reveals that the lines with limited/poor $\log(gf)$ quality mostly have $E_{low} > 4$ eV. We compile 25 electric dipole multiplets of 69 Fe I lines in BRASS, also shown in the right-hand panel of Figure 6. For each line of these multiplets we calculate the relative line strength ratios assuming single-configuration Russell–Saunders (LS) coupling and obeying the selection rules for these permitted transitions. The calculated multiplet line strengths (using Wigner 6j-symbol calculations) are normalized by scaling the LS-coupling $\log(gf)$-values to the strongest available principal line (marked with x_1 or x_2 in Figure 6), or the largest $\log(gf)$-value we calculate with the GRID method using the benchmark spectra. For a number of multiplets we find reasonable to good agreement between the LS and GRID relative $\log(gf)$-values. For example multiplet $a5P \to y5D$ ($E_{low} = 2.17$–2.23 eV) shows very similar distributions across its principal (x) and satellite (y and z) transitions. For multiplet $y3D \to y3F$ ($E_{low} = 4.73$–4.84 eV) the relative $\log(gf)$ distributions agree less, but also show differences between the literature and GRID $\log(gf)$-values of ~1.0 dex. This is also the case for $x5F \to 5F$ ($E_{low} = 4.9$–5.1 eV) and $y5D \to e5P$ ($E_{low} = 4.1$–4.23 eV) multiplets for which the relative LS and GRID $\log(gf)$-distributions across the x, y, and z line series are dissimilar and a re-normalization cannot remove the large differences above 0.5–1.0 dex for individual lines. The right-hand panel of Figure 6 shows that the differences between the normalized LS and the (best) BRASS GRID $\log(gf)$-values increase towards larger E_{low}-values.

For lines having $E_{low} > 4$ eV in the 25 studied multiplets the differences can increase above 0.5–1.0 dex mainly for the satellite transitions (marked y).

Using the interactive NIST Grotrian diagrams we find that both the $y5D$ and $x5F$ multiplet energy levels have level components that lie very close in energy to neighboring high energy levels of other atomic terms. For example the $y5D$ ($J = 1$) level at 4.217 eV falls next to another energy level at 4.220 eV corresponding to the $y5F$ ($J = 5$) level. In addition, the $x5F$ ($J = 2$) level at 5.085 eV lies near a $z5S$ ($J = 2$) level at 5.070 eV. The proximity of other nearby energy levels for these multiplet lower levels yields significant configuration interaction between the levels. The LS-coupling calculations cannot accurately predict the relative line strengths assuming single electronic configuration interaction. The increasing differences between the literature, LS and GRID $\log(gf)$-values towards larger E_{low} results from inaccurate theoretical Fe I $\log(gf)$-values due to poorly constrained configuration mixing coefficients and inaccurate or incomplete theoretical energy levels for the large number of close energy levels above $\sim$4 eV in the neutral Fe atom.

3. Summary

This paper presents new progress results for BRASS. We present new quality assessment results for the accuracy of atomic $\log(gf)$-values required for theoretical modeling of high-resolution stellar spectra using 7 FGK-type benchmark stars including the Sun. Astrophysical $\log(gf)$-values have been calculated for 1091 carefully selected sufficiently un-blended line transitions between 420 nm and 680 nm using two commonly employed methods. The agreement between both methods selects 845 lines suitable for our quality-assessments. An investigation of mean $\Delta\log(gf)$-values reveals large differences for lines with limited atomic data quality offered in the literature for $-3 \leq \log(gf) \leq -0.5$. We find that $\sim$53% of the quality-assessable lines have at least one literature $\log(gf)$-value in agreement with the astrophysical values, while values for other lines can differ by more than 0.5 dex. Only $\sim$38% of the investigated Fe I lines have sufficiently accurate literature $\log(gf)$-values, while $\sim$70–75% for other Fe-group element lines. The large percentage we find for theoretical Fe I $\log(gf)$-values of low quality offered in the literature mainly results from medium-strong and weak lines in multiplets with $E_{low} > 4$ eV, likely due to strong level mixing and inaccurate/incomplete energy levels. We also find that the majority of $\Delta\lambda$-values are below $\pm$0.01 Å, comparable to the high accuracy of the HERMES wavelength scale.

The Lines and Spectra BRASS Data Interface have been updated with the cross-matched atomic datasets and observed and theoretical stellar spectra. Users of the BRASS repository can query the LBDI for atomic data downloading, including the corresponding literature references, and the interactive display of dynamic plots for comparisons of database $\log(gf)$-values. The SBDI offers interactive display tools for the (observed and theoretical) benchmark spectra, combined with line identifications and atomic data values and line properties for user downloading. The BRASS repository offers interactive atomic data quality assessment pages for the 1091 investigated spectral lines. It also offers tools for interactive line equivalent width measurements and comprehensive help pages and tutorial videos to its users.

Author Contributions: BRASS conceptualization, A.L., P.R. and C.M.; methodology, A.L and M.L.; software development, A.L., M.L, and E.T.; results validation, A.L., M.L., P.R., M.D. and H.H.; project investigations, A.L., C.M., M.L., T.M., M.V.d.S. and P.A.M.v.H.; database resources, A.L.; data curation, A.L., M.L., C.M., T.M., and M.V.d.S.; writing—original draft preparation, A.L.; database visualization, A.L. and E.T.; supervision, A.L., M.D., H.H. and E.T.; project coordination, A.L.; project administration, A.L.; funding acquisition, A.L., P.R. and C.M.

Funding: The research for the present results has been subsidised by the Belgian Federal Science policy Office under contract No. BR/143/A2/BRASS.

Acknowledgments: The BRASS Team would like to thank everybody who directly or indirectly has contributed since 2015 to the realization of the BRASS project.

Conflicts of Interest: The authors declare no conflict of interest.

Abbreviations

The following abbreviations are used in this manuscript:

LBDI	Lines BRASS Data Interface
SBDI	Spectra BRASS Data Interface
NIST	National Institute of Standards and Technology
VALD	Vienna Atomic Line Database
VAMDC	Virtual Atomic and Molecular Data Centre
TIPbase	The Iron Project database
TOPbase	The Opacity Project database
CHIANTI	The CHIANTI atomic database
SpectroWeb	The SpectroWeb database
Spectr-W^3	The Spectr-W^3 database
HERMES	High Efficiency and Resolution Mercator Echelle Spectrograph
FTS	Fourier Transform Spectrograph
SNR	Signal to Noise Ratio
FWHM	Full-Width Half-Maximum

References

1. Neckel, H.; Labs, D. The solar radiation between 3300 and 12500 Å. *Sol. Phys.* **1984**, *90*, 205. [CrossRef]
2. Raskin, G.; Van Winckel, H.; Hensberge, H.; Jorissen, A.; Lehmann, H.; Waelkens, C.; Avila, G.; De Cuyper, J.P.; Degroote, P.; Dubosson, R.; et al. HERMES: A high-resolution fibre-fed spectrograph for the Mercator telescope. *Astron. Astrophys.* **2011**, *526*, A69. [CrossRef]
3. Lobel, A.; Royer, P.; Martayan, C.; Laverick, M.; Merle, T.; David, M.; Hensberge, H.; Thienpont, E. The Belgian repository of fundamental atomic data and stellar spectra. *Can. J. Phys.* **2017**. [CrossRef]
4. Lobel, A. Oscillator Strength Measurements of Atomic Absorption Lines from Stellar Spectra. *Can. J. Phys.* **2011**, *89*, 395–402. [CrossRef]
5. Merle, T.; Laverick, M.; Lobel, A.; Royer, P.; Van der Swaelmen, M.; Frémat, Y.; Sekaran, S.; Martayan, C.; van Hoof, P.; David, M.; et al. The Belgian Repository of fundamental Atomic data and Stellar Spectra (BRASS). In *Proc. of Semaine de l'Astrophysique Francaise 2018*; Di Matteo, P., Billebaud, F., Herpin, F., Lagarde, N., Marquette, J.-B., Robin, A., Venot, O., Eds.; Société Francaise d'Astronomie et d'Astrophysique (SF2A): Bordeaux, France, 2018; p. 153.
6. Laverick, M.; Lobel, A.; Merle, T.; Royer, P.; Martayan, C.; David, M.; Hensberge, H.; Thienpont, E. The Belgian repository of fundamental atomic data and stellar spectra (BRASS) I. Cross-matching atomic databases of astrophysical interest. *Astron. Astrophys.* **2018**, *612*, A60. [CrossRef]
7. Laverick, M.; Lobel, A.; Royer, P.; Merle, T.; Martayan, C.; van Hoof, P.; Van der Swaelmen, M.; David, M.; Hensberge, H.; Thienpont, E. The Belgian repository of fundamental atomic data and stellar spectra (BRASS) II. Quality assessment of atomic data for unblended lines in FGK stars. *Astron. Astrophys.* **2019**, *624*, A60. [CrossRef]
8. Lobel, A.; Royer, P.; Martayan, C.; Laverick, M.; Merle, T.; van Hoof, P.A.M.; Van der Swaelmen, M.; David, M.; Hensberge, H.; Thienpont, E. The Belgian Repository of Fundamental Atomic Data and Stellar Spectra: Atomic Line Data Validation. In *Workshop on Astrophysical Opacities*; Mendoza, C., Turck-Chiéze, S., Colgan, J., Eds.; Astron. Soc. of the Pacific CS: San Francisco, CA, USA, 2018; Volume 515, pp. 255–262.

Article

Coulomb (Velocity) Gauge Recommended in Multiconfiguration Calculations of Transition Data Involving Rydberg Series

Asimina Papoulia [1,2,*], Jörgen Ekman [1], Gediminas Gaigalas [3], Michel Godefroid [4], Stefan Gustafsson [1], Henrik Hartman [1], Wenxian Li [1], Laima Radžiūtė [3], Pavel Rynkun [3], Sacha Schiffmann [2,4], Kai Wang [5] and Per Jönsson [1]

[1] Department of Materials Science and Applied Mathematics, Malmö University, SE-20506 Malmö, Sweden; jorgen.ekman@mau.se (J.E.); stefan.gustafsson@mau.se (S.G.); henrik.hartman@mau.se (H.H.); wenxian.li@mah.se (W.L.); per.jonsson@mau.se (P.J.)

[2] Division of Mathematical Physics, Department of Physics, Lund University, SE-22100 Lund, Sweden; saschiff@ulb.ac.be

[3] Institute of Theoretical Physics and Astronomy, Vilnius University, Saulėtekio av. 3, LT-10222 Vilnius, Lithuania; gediminas.gaigalas@tfai.vu.lt (G.G.); laima.radziute@tfai.vu.lt (L.R.); pavel.rynkun@tfai.vu.lt (P.R.)

[4] Chimie Quantique et Photophysique, Université libre de Bruxelles, B-1050 Brussels, Belgium; michel.godefroid@ulb.be

[5] Hebei Key Lab of Optic-electronic Information Materials, College of Physics Science and Technology, Hebei University, Baoding 071002, China; kaiwang1128@aliyun.com

[*] Correspondence: asimina.papoulia@mau.se; Tel.: +46-40-66-58268

Received: 23 October 2019; Accepted: 21 November 2019; Published: 26 November 2019

Abstract: Astronomical spectroscopy has recently expanded into the near-infrared (nIR) wavelength region, raising the demands on atomic transition data. The interpretation of the observed spectra largely relies on theoretical results, and progress towards the production of accurate theoretical data must continuously be made. Spectrum calculations that target multiple atomic states at the same time are by no means trivial. Further, numerous atomic systems involve Rydberg series, which are associated with additional difficulties. In this work, we demonstrate how the challenges in the computations of Rydberg series can be handled in large-scale multiconfiguration Dirac–Hartree–Fock (MCDHF) and relativistic configuration interaction (RCI) calculations. By paying special attention to the construction of the radial orbital basis that builds the atomic state functions, transition data that are weakly sensitive to the choice of gauge can be obtained. Additionally, we show that the Babushkin gauge should not always be considered as the preferred gauge, and that, in the computations of transition data involving Rydberg series, the Coulomb gauge could be more appropriate for the analysis of astrophysical spectra. To illustrate the above, results from computations of transitions involving Rydberg series in the astrophysically important C IV and C III ions are presented and analyzed.

Keywords: infrared spectra; spectrum calculations; multiconfiguration methods; Rydberg series; Rydberg states; electric dipole transitions; transition rates; Babushkin gauge; Coulomb gauge; length form; velocity form

1. Introduction

The starlight emitted at optical or shorter wavelengths is efficiently scattered by intervening interstellar and intergalactic dust particles. To observe stars deeper into the galactic center and go even beyond the Milky Way, astrophysical missions and spectrographs were recently designed to observe nIR radiation, which has higher transmission through dust clouds [1–3]. Accurate transition data from the IR part of the spectrum are thus required to interpret the spectra of distant astronomical objects observed, and to carry out chemical abundance studies.

The interest in the nIR region is relatively recent, and atomic data corresponding to wavelengths from 1 to 5 μm are scarce. Due to the limited resources and the numerous possible transitions, laboratory measurements are insufficient to provide astrophysicists with complete sets of atomic transition data. Critically evaluated theoretical data are, therefore, necessary to complement experiments and to allow for accurate chemical abundance analyses of stars. In the long wavelength IR regime, lines of atoms are produced by transitions between states lying close in energy, which often correspond to transitions between highly excited states. The latter instance necessitates atomic structure calculations over a large portion of a spectrum. Extensive spectrum calculations that produced transition data in the nIR region were formerly carried out as part of the Opacity Project [4]. The latter non-relativistic calculations were based on the close-coupling approximation of the R-matrix theory.

Performing spectrum calculations, in which multiple atomic states are targeted at the same time, is generally not trivial. In multiconfiguration calculations, the correlation between the electrons is taken into account by expanding the targeted states in a number of symmetry adapted basis functions, which are built from products of spin-orbitals. To accurately predict the energies of all the targeted states, the shapes of the radial parts of the spin-orbitals must be such that they account for the LS-term dependencies; i.e., the way the electrons are coupled to form different terms from the same configuration [5]. Additionally, many studies involve states that are part of Rydberg series. Perturbers often enter the Rydberg series and the atomic state expansions must correctly predict their positions [6]. Computations of Rydberg series have to further describe states with electron distributions occupying different regions in space, extending far out from the atomic core. The above challenges require that special attention is paid to the optimization scheme of the wave functions; i.e., how the orbital basis is generated. The challenges in the computations of Rydberg series become more apparent when computing transition data.

The transition parameters (line strengths, oscillator strengths, and transition rates) are expressed in terms of reduced matrix elements of the transition operator. Different choices of gauge, Babushkin and Coulomb, for the transition operator lead to alternative expressions for the reduced matrix elements, and consequently, the transition parameters. Gauge invariance of the transition data is a straightforward matter for hydrogenic systems. Yet, the use of approximate wave functions results in different values for transition data expressed in different gauges. During the past years, recommendations for choosing the appropriate gauge became contradictory, suggesting further work in the field [7–12]. The Babushkin gauge (or length form) is sensitive to the outer part of the wave functions that governs the atomic transitions, and transition data expressed in this gauge are often considered to be more reliable than transition data expressed in the Coulomb gauge (or velocity form) [13]. It is, however, argued that provided reasonably accurate approximate wave functions, the Coulomb gauge (or velocity form) may give the best results when the transition energy is not very small [14]. Recent work suggests that the Coulomb gauge gives more accurate results and is the preferred gauge for transitions involving high Rydberg states [15].

In this paper, we present and analyze results from computations of Rydberg series in the C IV and C III ions. Although the latter are of astrophysical interest, the goal of the paper is not benchmarking transition data for these two ions against other theoretical methods, but instead assessing the relative reliability of the MCDHF/RCI results obtained with the two different gauges. Using the MCDHF method, we apply different computational strategies for optimizing the radial orbital basis used for constructing the wave functions and compare the results. For transitions involving low-lying states, the transition data are accurately computed in both the Babushkin and the Coulomb gauge, independently of how the radial orbitals are optimized. On the other hand, transitions involving high Rydberg states are problematic, and the Babushkin gauge does not provide trustworthy results when conventional optimization strategies are applied. However, by paying special attention to the construction of the radial orbital basis that builds the atomic state functions, transition data that are weakly sensitive to the choice of gauge are produced for all the computed transitions in the ions we study. The present article is an extended transcript of the poster presentation given on 24 June 2019 at the 13th International Colloquium on Atomic Spectra and Oscillator Strengths (ASOS2019) for Astrophysical and Laboratory Plasmas that took place at Fudan University in Shanghai, China (https://asos2019.fudan.edu.cn).

2. Theory

2.1. MultiConfiguration Calculations

Numerical representations of atomic state functions (ASFs), which are approximations to the exact wave functions, are obtained using the fully relativistic MCDHF method [16,17]. In the MCDHF method, the ASFs $\Psi(\gamma \pi J M_J)$ are expanded over N_{CSF} antisymmetrized basis functions $\Phi(\gamma_\nu \pi J M_J)$, which are known as configuration state functions (CSFs), i.e.,

$$\Psi(\gamma \pi J M_J) = \sum_{\nu=1}^{N_{CSF}} c_\nu \Phi(\gamma_\nu \pi J M_J). \tag{1}$$

In the expression above, J and M_J are the angular momentum quantum numbers, π is the parity, and γ_ν denotes other appropriate labeling of the CSF ν, such as orbital occupancy and angular coupling tree. The CSFs are coupled products of one-electron Dirac orbitals $\psi_{n\kappa,m}$, which have the general form:

$$\psi_{n\kappa,m}(\mathbf{r}) = \frac{1}{r} \begin{pmatrix} P_{n\kappa}(r)\chi_{\kappa,m}(\theta,\varphi) \\ iQ_{n\kappa}(r)\chi_{-\kappa,m}(\theta,\varphi) \end{pmatrix}, \tag{2}$$

where $\chi_{\pm\kappa,m}(\theta,\varphi)$ are the two-component spin-angular functions and $\{P_{n\kappa}(r), Q_{n\kappa}(r)\}$ are, respectively, the radial functions of the large and small components, which are represented on a logarithmic grid. The selection of the CSFs to be included in the expansion (1) depends on the shell structure of the atom at hand and the computed properties, as explained in Section 3. The shape of the radial functions $\{P_{n\kappa}(r), Q_{n\kappa}(r)\}$ is determined by the effective field in which the considered electron moves, which is in turn established by the included CSFs [18].

The expansion coefficients c_ν, together with the radial parts of the spin-orbitals, are obtained in a self-consistent field (SCF) procedure. The set of SCF equations to be iteratively solved results from applying the variational principle on a weighted energy functional of all the targeted atomic states according to the extended optimal level (EOL) scheme [19]. In fully relativistic calculations, the energy functional is estimated from the expectation value of the Dirac-Coulomb Hamiltonian [17]. The angular integrations needed for the construction of the energy functional are based on the second quantization formalism in the coupled tensorial form [20,21].

The MCDHF method is employed to generate an orbital basis. Given this basis, the final wave functions $\Psi(\gamma \pi J M_J)$ of the targeted states are determined in subsequent RCI calculations. In the RCI calculations, the spin-orbitals defining the basis are fixed and only the expansion coefficients c_ν are evaluated by diagonalizing the Hamiltonian matrix. At this step, the expansions based on Equation (1) can be augmented to include CSFs that account for additional electron correlation effects. All MCDHF and RCI calculations were performed using the relativistic atomic structure package GRASP2018 [22].

2.2. Transition Parameters

Once the wave functions $\Psi(\gamma \pi J M_J)$ have been determined, transition parameters can be computed. In this work, we focus on the computation of transition rates (or probabilities) for electric dipole (E1) transitions. Electric dipole transitions are much stronger than electric quadrupole (E2) and magnetic multipole (Mk) transitions. For the transition rate $A^{(k)}$ of electric dipole ($k = 1$) emission from an upper state $\gamma' \pi' J' M'_J$ to any of the $2J + 1$ states $\gamma \pi J M_J$ of lower energy, we have the following proportionality

$$A^{(1)}(\gamma' \pi' J', \gamma \pi J) \sim (E_{\gamma' \pi' J'} - E_{\gamma \pi J})^3 \, \frac{S^{(1)}(\gamma \pi J, \gamma' \pi' J')}{2J' + 1}, \tag{3}$$

where $E_{\gamma' \pi' J'} - E_{\gamma \pi J}$ is the transition energy and $S^{(1)}(\gamma \pi J, \gamma' \pi' J')$ is the line strength given by

$$S^{(1)}(\gamma \pi J, \gamma' \pi' J') = |\langle \Psi(\gamma \pi J) || \mathbf{O}^{(1)} || \Psi(\gamma' \pi' J') \rangle|^2. \tag{4}$$

The E1 transition rates are therefore expressed in terms of reduced matrix elements of the electric dipole transition operator $\mathbf{O}^{(1)}$. From Equation (1), it follows that

$$\langle \Psi(\gamma \pi J) || \mathbf{O}^{(1)} || \Psi(\gamma' \pi' J') \rangle = \sum_{k,l} c_k c'_l \langle \Phi(\gamma_k \pi J) || \mathbf{O}^{(1)} || \Phi(\gamma'_l \pi' J') \rangle. \tag{5}$$

The choice of gauge parameter determines whether the electric dipole matrix elements are computed in the Babushkin or the Coulomb gauge, which in non-relativistic calculations correspond to the length and the velocity form, respectively [10]. The two forms are equivalent for hydrogenic wave functions, but they result in different values when approximate many-electron wave functions are used. As shown later, in Section 4, these values reveal a strong dependence on the generated orbital basis and the captured correlation effects. Although the present results arise from fully relativistic calculations, similar behavior is observed when non-relativistic multiconfiguration calculations are performed [15].

The explicit expressions of the electric dipole reduced matrix elements in the Babushkin and the Coulomb gauge are given in [10]. Taking for convenience the non-relativistic limit, the electric dipole reduced matrix elements are, in the length and the velocity form, respectively, given by

$$\langle \Psi(\gamma \pi J) || \sum_{i=1}^{N} r_i \mathbf{C}^{(1)}(i) || \Psi(\gamma' \pi' J') \rangle \tag{6}$$

and

$$\frac{1}{E_{\gamma' \pi' J'} - E_{\gamma \pi J}} \langle \Psi(\gamma \pi J) || \sum_{i=1}^{N} \nabla^{(1)}(i) || \Psi(\gamma' \pi' J') \rangle, \tag{7}$$

where the summation runs over the number N of electrons and $\mathbf{C}^{(1)}$ is the renormalized spherical harmonic of rank 1 [23]. The reduced matrix elements of (6) and (7) involve, respectively, sums over radial transition integrals of the kind

$$\int_0^\infty P(r) r P'(r) \, dr \tag{8}$$

and

$$\int_0^\infty P(r) \frac{d}{dr} P'(r) \, dr, \tag{9}$$

weighted with the products of the expansion coefficients of the CSFs and the angular coefficients [20,21,23]. $P'(r)$ and $P(r)$ are the radial functions of the large components of the Dirac one-electron spin-orbitals (2) that build the CSFs of the initial state $\gamma' \pi' J'$ and the final state $\gamma \pi J$, respectively. In the present work, the initial and final states belonging to different parities are built from a common orbital basis.

In the computation of the integrals (8), the emphasis is given to the tail of the radial orbitals, while in the integrals (9) the emphasis is given instead to the inner part of the radial orbitals. In the simple Hartree–Fock (HF) model, the approximate wave functions usually display a correct asymptotic behavior towards large r (see also Section 5), and since the former integrals are also computationally simpler, the transition rates are traditionally provided in the length form [7–9,14]. As discussed in

Section 5, when multiconfiguration methods concurrently target multiple atomic states, all wave functions are not always well approximated at large r; and the velocity form, or correspondingly, the Coulomb gauge, may by contrast, give the best results.

The agreement between the transition rates A_B and A_C, respectively, evaluated in the Babushkin and the Coulomb gauge is used as an indicator of accuracy. This is particularly useful when laboratory measurements are not available for comparison. The uncertainty of the computed transition rates in the preferred gauge can be estimated as

$$dT = \frac{|A_B - A_C|}{\max(A_B, A_C)},$$ (10)

which reflects the relative discrepancy between the Babushkin and the Coulomb gauge of the computed line strengths [24,25]. Although the accuracy indicators dT should be used in a statistical manner for a group of transitions with similar properties (see [6]), individual dT values can point out problematic transitions, which could further be analyzed.

3. Computational Methodology—Optimization of the Orbital Basis

The accuracy of multiconfiguration calculations relies on the CSF expansion of Equation (1). A first approximation of the ASFs is acquired by performing an MCDHF calculation on expansions that are built from the configurations that define what is known as the multi-reference (MR) [17]. The orbitals that take part in this initial calculation are called spectroscopic orbitals and are kept frozen in all subsequent calculations. The initial approximation of the ASFs is improved by augmenting the expansion with CSFs that interact with the ones that are generated by the MR configurations. Such CSFs are built from configurations that differ by either a single (S) or a double (D) electron substitution from the configurations in the MR [17,26]. Following the SD-MR scheme, the interacting configurations are obtained by allowing substitutions of electrons from the spectroscopic orbitals to an active set of correlation orbitals, which is systematically increased (each step introducing an additional correlation orbital layer) [27,28]. These configurations produce CSFs that can be classified, based on the nature of the substitutions, into CSFs that capture valence–valence (VV), core–valence (CV), and core–core (CC) electron correlation effects.

Building accurate wave functions requires a very large orbital basis. Even so, a large but incomplete orbital basis does not ensure that the wave functions give accurate properties other than energies. In the MCDHF calculations, the correlation orbitals are obtained by applying the variational principle on the weighted energy functional of all the targeted atomic states. Thus, the orbitals of the first correlation layers will overlap with the spectroscopic orbitals that account for the effects that minimize the energy the most [18,29]. The energetically dominant effects must first be saturated to obtain orbitals localized in other regions of space, which might describe effects that do not lower the energy much, but are important for, e.g., transition parameters. One must, therefore, carefully choose the orbital basis with respect to the computed properties [30].

Valence atomic transitions are governed by the outer part of the wave functions and this part must be properly described by the correlation orbitals to obtain reliable transition parameters. States that are part of Rydberg series encompass valence orbitals of increasing principal quantum number n. Spectrum calculations that involve Rydberg series need, thus, to describe states with electron distributions localized in different regions of space extending far out from the atomic core. Since the overlap between orbitals describing Rydberg states is in some cases minor, generating an optimal orbital basis is not straightforward [6]. This raises the need to explore different computational strategies.

3.1. C IV

In lithium-like carbon, the configurations being studied are $1s^2 nl$ with $n = 2$ to 8 and $l = 0$ to 4 and $1s^2 6h$. These configurations define the MR and correspond to 53 targeted atomic states of both even and odd parity, which are simultaneously optimized. For simple systems such as three-electron systems, the MCDHF calculations are conventionally performed using CSF expansions that are produced by SD-MR electron substitutions from all spectroscopic orbitals. In this manner, the CSFs capture all valence (V), CV, and CC correlation effects. The $1s1s$ pair-correlation effect is energetically very important and the orbitals of the first correlation layers overlap with the $1s$ core orbital accounting for this effect (see Table 1). After building six correlation layers by utilizing this conventional approach, we see that all correlation orbitals up to $14s, 14p, 14d, 12f, 12g, 8h$, and $7i$ are rather contracted in comparison with the outer Rydberg orbitals. As a consequence, the wave functions are not properly described for all states, and in particular, not for the higher Rydberg states considered.

Table 1. The mean radii $\langle r \rangle$ (a.u.) of the spectroscopic and correlation orbitals that belong to the s and p symmetries in C IV. The correlation orbitals result from two different optimization schemes, the conventional and the alternative, and they occupy different regions in space.

Spectroscopic			Correlation		Spectroscopic			Correlation	
			Conventional	Alternative				Conventional	Alternative
$1s$	0.27	$9s$	0.51	1.12			$9p$	0.44	0.87
$2s$	1.31	$10s$	0.43	0.92	$2p$	1.28	$10p$	0.41	1.03
$3s$	3.00	$11s$	0.42	0.84	$3p$	2.95	$11p$	0.40	1.00
$4s$	5.55	$12s$	0.46	0.87	$4p$	5.64	$12p$	0.45	1.18
$5s$	8.81	$13s$	0.56	0.87	$5p$	8.99	$13p$	0.48	2.59
$6s$	12.82	$14s$	0.40	1.26	$6p$	13.10	$14p$	0.77	5.94
$7s$	17.58				$7p$	17.94			
$8s$	23.09				$8p$	23.54			

For a more appropriate description of the wave functions, the correlation orbitals must occupy the space between the $1s$ core orbital and the inner valence orbitals. This can be accomplished by imposing restrictions on the allowed substitutions for obtaining the orbital basis. Thus, the MCDHF calculations are alternatively performed using CSF expansions that are produced by SD-MR substitutions with the restriction of allowing maximum one hole in the $1s$ core shell. In this case, the shape of the correlation orbitals is established by CSFs accounting for V and CV correlation effects. The resulting correlation orbitals are, as shown in Table 1, more extended, overlapping with orbitals of higher Rydberg states.

The final wave functions of the targeted states are determined in subsequent RCI calculations, where D substitutions from the $1s$ core orbital and triple (T) substitutions from all the spectroscopic orbitals, are included. The number of CSFs in the final even and odd state expansions are, respectively, 1,077,872 and 1,287,706, distributed over the different J symmetries.

3.2. C III

In beryllium-like carbon, the configurations in question are $1s^22snl$ with $n = 2$ to 7 and $l = 0$ to 4 and $1s^22p^2$, $1s^22p3s$, $1s^22p3p$, and $1s^22p3d$. These configurations define the MR and correspond to 114 targeted atomic states of both even and odd parity, which are simultaneously optimized. Having introduced two correlation orbitals, $8s$ and $8p$—specifically targeted to account for the LS-term dependence [31], i.e., the difference between the ns orbitals for $2sns\ ^3S$ and $2sns\ ^1S$ and the difference between the np orbitals for $2snp\ ^3P°$ and $2snp\ ^1P°$—the MCDHF calculations are conventionally performed using CSF expansions that are produced by SD-MR electron substitutions from all spectroscopic orbitals with the restriction that only one excitation is allowed from the $1s^2$ atomic core. In this manner, the CSFs capture VV and CV correlation effects. The $1snl$ pair-correlation effect is comparatively important, and the orbitals of the first correlation layers are spatially localized between the $1s$ orbital and the $2s$ and $2p$ orbitals. As the CV correlation effects start to saturate, the correlation orbitals are gradually located further away from the $1s^2$ atomic core (see Table 2). The correlation orbitals up to $12s, 12p, 12d, 12f, 11g$, and $8h$ are, however, still contracted in comparison with the outer Rydberg orbitals.

Table 2. Same as Table 1, but for radial orbitals in C III. The correlation orbitals 8s and 8p, which are introduced to account for the *LS*-term dependencies, are the same in both optimization schemes and fairly diffuse in comparison with the rest of the correlation orbitals.

Spectroscopic		Correlation		Spectroscopic			Correlation	
		Conventional	Alternative				Conventional	Alternative
1s	0.26	9s	1.05	4.86		9p	1.00	3.56
2s	1.28	10s	1.48	3.67	2p	1.23	10p	1.24	3.37
3s	3.57	11s	1.88	3.25	3p	3.74	11p	1.56	3.37
4s	6.63	12s	1.87	8.40	4p	7.04	12p	1.52	9.04
5s	10.80				5p	11.37			
6s	15.95				6p	16.71			
7s	22.10				7p	23.04			
term corr.					**term corr.**				
8s	8.22				8p	5.55			

For a more appropriate description of the wave functions, the correlation orbitals must occupy the space of the valence orbitals. In the alternative approach, this is accomplished by allowing SD substitutions only from the outer valence orbitals accounting for VV correlation. The resulting correlation orbitals are, as shown in Table 2, more extended, overlapping with orbitals of higher Rydberg states.

The final wave functions of the targeted states are determined in subsequent RCI calculations, where SDT substitutions from all the spectroscopic orbitals are included, with the restriction that only one substitution is allowed from the $1s^2$ atomic core. The number of CSFs in the final even and odd state expansions are, respectively, 1,578,620 and 1,274,147, distributed over the different *J* symmetries.

4. Results

Excitation energies are produced, based on the conventional and alternative computational strategies that were described in Section 3, and are compared with the critically evaluated data from the National Institute of Standards and Technology's (NIST's) Atomic Spectra Database (ASD) [32]. In C IV, the computed excitation energies are in excellent agreement with the NIST's recommended values. Both computational approaches give similar energies and the relative differences from the NIST values are less than 0.01%. For the more complex system of C III, the computed excitation energies agree also well with the energies proposed by NIST. The relative differences between theoretical and critically compiled energies are, on average, of the order of 0.1% and 0.02%, when following the conventional and the alternative approach, respectively. The NIST database does not provide excitation energies for the $2s6s\ ^2S$, $2s7s\ ^2S$, and $2s7p\ ^3P^\circ$ states, which are included in the computations.

Transition rates *A* are produced based on the two different computational strategies. In the present computations, the uncertainties in the predicted excitation energies of two states associated with a transition most often cancel out, and consequently, the majority of the evaluated transition energies are ultimately in perfect agreement with the NIST values. The uncertainties of the computed transition rates solely emerge from the disagreement of the computed line strengths in the Babushkin and the Coulomb gauge, which are then reflected in the dT values. When the conventional strategy is applied, most of the transition rates are predicted with uncertainties dT lower than 1% and 5%, in lithium-like and beryllium-like carbon, respectively. Yet, for transitions involving high Rydberg states, the uncertainties increase remarkably, especially for the more complex C III ion. The alternative strategy for optimizing the radial orbitals yields transition rates that are overall more accurate. The improvement in accuracy is significant for transitions that involve high Rydberg states.

The uncertainties dT of the transition rates computed with the conventional and alternative approaches are presented and analyzed for groups of transitions in the studied carbon ions. Each group is selected to include transitions between a fixed state and Rydberg states described by electron distributions that are gradually localized farther from the atomic core. Accordingly, Figure 1a,b illustrates the uncertainties dT for the $2p\,^2P^\circ_{1/2} - ns\,^2S_{1/2}$ and $np\,^2P^\circ_{1/2} - 8s\,^2S_{1/2}$ groups of transitions in C IV. Similarly, Figure 2a,b illustrates the dT values for the $2s^2\,^1S_0 - 2snp\,^1P^\circ_1$ and $2sns\,^1S_0 - 2s7p\,^1P^\circ_1$ groups of transitions in C III.

Figure 1a demonstrates the uncertainties for the series of transitions between the low-lying $2p\,^2P^\circ_{1/2}$ state and successively higher Rydberg $ns\,^2S_{1/2}$ states. The uncertainty dT of the transition rates computed with the conventional approach grows almost exponentially with the increasing principal quantum number n. The transition rate for the $2p\,^2P^\circ_{1/2} - 8s\,^2S_{1/2}$ transition, which is the transition between the two states with the largest energy difference in the plot, eventually exhibits the highest uncertainty—12.4%. When the alternative approach is utilized instead, the uncertainties range between 0% and 0.4% for the respective transitions. The same trends are also observed in other groups of transitions in C IV, such as the $2p\,^2P^\circ - nd\,^2D$, the $2s\,^2S - np\,^2P^\circ$, and so forth.

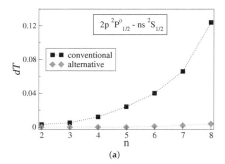 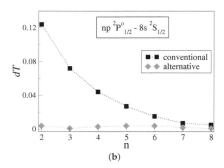

(a)   (b)

Figure 1. (**a**) The uncertainty dT of the computed transition rates for transitions between the $2p\,^2P^\circ_{1/2}$ state and Rydberg $ns\,^2S_{1/2}$ states of increasing principal quantum number n in C IV. The black squares and magenta diamonds, respectively, correspond to the results from the conventional and the alternative strategies for optimizing the radial orbitals. (**b**) Same as the first panel, but for transitions between the $8s\,^2S_{1/2}$ state and successive Rydberg $np\,^2P^\circ_{1/2}$ states in C IV.

Having as a starting point the $2p\,^2P^\circ_{1/2} - 8s\,^2S_{1/2}$ transition, Figure 1b demonstrates the uncertainties for the series of transitions between the high Rydberg $8s\,^2S_{1/2}$ state and successively higher Rydberg $np\,^2P^\circ_{1/2}$ states. As n increases, the transition energy gets smaller. The uncertainties of the transition rates computed with the conventional approach exhibit a nearly exponential decay with increasing n. Similarly to Figure 1a, when following the alternative strategy, the uncertainties in the transition rates are substantially reduced, ranging between 0.1% and 0.4%. Other groups of transitions in C IV, such as the $np\,^2P^\circ - 8d\,^2D$ series and the $ns\,^2S - 8p\,^2P^\circ$ series, follow analogous trends.

Figure 2a displays the dT values for transitions between the low-lying $2s^2\,^1S_0$ state and successively higher Rydberg $2snp\,^1P^\circ_1$ states. Looking at Figure 2a, when the conventional approach is applied the uncertainties dT increase sharply for $n > 5$. For the $2s^2\,^1S_0 - 2s7p\,^1P^\circ_1$ transition, i.e., the transition between the two states with the largest energy difference, the dT rises to 63%. The latter is about five times larger than the highest estimated dT value in the figures above. Once more, when the radial orbitals are optimized using the alternative strategy, the uncertainties drop dramatically, ranging between 0% and 1.4% for the respective transitions. More groups of transitions in C III that reveal similar behavior are the $2s2p\,^1P^\circ_1 - 2sns\,^1S_0$ series and the $2p^2\,^3P_0 - 2snp\,^3P^\circ_1$ series.

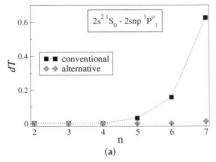
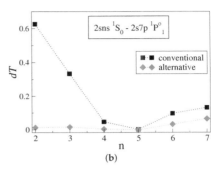

(a) (b)

Figure 2. (a) The uncertainty dT of the computed transition rates for transitions between the $2s^2\ {}^1S_0$ state and Rydberg $2snp\ {}^1P_1^\circ$ states of increasing principal quantum number n in C III. The black squares and magenta diamonds, respectively, correspond to the results from the conventional and alternative strategies for optimizing the radial orbitals. (b) Same as the first panel, but for transitions between the $2s7p\ {}^1P_1^\circ$ state and successive Rydberg $2sns\ {}^1S_0$ states in C III.

Starting with the $2s^2\ {}^1S_0 - 2s7p\ {}^1P_1^\circ$ transition, Figure 2b displays the dT values for transitions between the high Rydberg $2s7p\ {}^1P_1^\circ$ state and successively higher Rydberg $2sns\ {}^1S_0$ states. Likewise, in Figure 1b, the increase in n corresponds to transitions between states that gradually come closer in energy. The uncertainties of the transition rates computed with the conventional approach reduce rapidly as n increases. Applying the alternative strategy results in much lower uncertainties, which extend between 0.5% and 6.6%. A similar trend is also observed in the $2snp\ {}^1P_1^\circ - 2s7s\ {}^1S_0$ series of transitions in C III. Although the last two points in Figure 2b correspond to transitions between states lying close in energy, the uncertainties are comparatively high. Nevertheless, the alternative strategy still predicts the transition rates with lower uncertainties.

Altogether, for transitions between low-lying states, the transition rates are accurately predicted independently of whether the conventional or the alternative computational strategy is employed. Further, when both states involved in a transition in C IV are high Rydberg states, the transition rates are also predicted with high accuracy in both computations. On average, the same holds for transitions between high Rydberg states in C III. The line strengths of transitions between two states close in energy, and with the outer electrons occupying nearly the same region of space, are relatively large, and therefore, weakly affected by the optimization strategy of the radial orbitals. Quite the contrary, transitions between a low-lying state and a high Rydberg state are problematic in both carbon ions. The line strength of transitions between two states with large energy differences, and with the outer electrons occupying different parts of space, take smaller values, which are more sensitive to how the radial orbitals are optimized with regard to correlation.

To better understand the origins of the large dT values in transitions between low-lying states and high Rydberg states, the convergences of the individual transition rates A_B and A_C, computed in the Babushkin and the Coulomb gauges respectively, are studied with respect to the increasing active set of correlation orbitals. In connection with the figures above, this is done for the $2p\ {}^2P_{1/2}^\circ - 8s\ {}^2S_{1/2}$ transition in C IV (see Figure 3a) and $2s^2\ {}^1S_0 - 2s7p\ {}^1P_1^\circ$ transition in C III (see Figure 3b); i.e., the transitions with the highest uncertainties dT. In Figure 3a,b, the convergences of the A_B and A_C values are illustrated for the two different computational approaches.

As seen in Figure 3a,b, when the computations are performed in the alternative manner, the transition rates given by A_B and A_C ultimately come really close in value. Considering the small final dT value, the agreement between the A_B and A_C values is expected. One observes that the transition rate given by A_C is rather stable with respect to the increasing orbital set. The A_C value varies by only 1% and 6.2% for each of the transitions displayed in the figures below. On the contrary, the A_B value varies by 13.7% and 27.2%, respectively. In the $2p\,^2P^{\circ}_{1/2} - 8s\,^2S_{1/2}$ transition, it takes five correlation layers for the A_B value to start converging, while in the $2s^2\,^1S_0 - 2s7p\,^1P^{\circ}_1$ transition it takes three layers of correlation orbitals for the A_B value to converge.

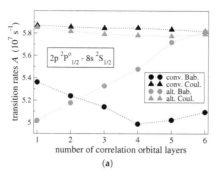 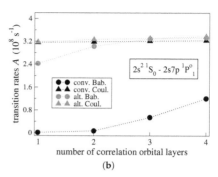

(a) (b)

Figure 3. (**a**) The transition rates A in the Babushkin (circles) and the Coulomb (triangles) gauges for the $2p\,^2P^{\circ}_{1/2} - 8s\,^2S_{1/2}$ transition in C IV, as a function of the increasing number of correlation layers. The transition rates computed in the conventional and the alternative manner are respectively shown in black and magenta. (**b**) Same as the first panel, but for the $2s^2\,^1S_0 - 2s7p\,^1P^{\circ}_1$ transition in C III.

Looking at Figure 3a,b, when the conventional approach is applied, the individual A_B and A_C values do not converge, as the large final uncertainties dT reveal. The transition rate given by A_C is, however, again stable and is also consistent with the A_B and A_C values provided by the alternative computational strategy. Throughout the optimization of the radial orbitals in the conventional manner, the A_C value varies by only 0.9% and 2.3% for each of the transitions displayed in Figure 3a,b, respectively. Although it seems that the A_B values will eventually approach the A_C ones, this would require a very large orbital basis, which is beyond the reach of the available computational resources. One may deduce that when the conventional computational strategy is applied the transition rates A_B, computed in the Babushkin gauge, are problematic and unreliable.

5. Discussion

Transition data, such as transition rates A, are expressed in terms of reduced matrix elements of the transition operator (see Equation (5)), which can be computed in different gauges. According to Equation (10), the uncertainty of the computed A values is assessed by the agreement of the transition rates computed in the different gauges. Computations of reduced matrix elements in different gauges, however, probe separate parts of the wave functions. Hence, the radial parts of the wave functions must be well approximated as a whole to obtain gauge invariant transition rates.

For transitions between low-lying states, both computational strategies yield reduced matrix elements of the transition operator that almost reach gauge invariance, and the transition rates are, overall, accurately predicted. There are enough correlation orbitals spatially localized between the core and the inner valence orbitals that make up the low-lying states. As a result, the inner parts of the wave functions are adequately approximated. Moreover, the spectroscopic outer valence orbitals, which make up the higher Rydberg states and are localized farther from the atomic core, improve the description of the outer parts of the wave functions for representing the low-lying states, ensuring that they have the correct asymptotic behavior. The radial parts of the latter wave functions are then effectively described at all r values, being insensitive to the choice of the optimization strategy with regard to correlation.

For transitions between a low-lying state and a high Rydberg state, the conventional computational strategy fails to produce accurate transition rates. The correlation orbitals are significantly contracted compared to the outer Rydberg orbitals. Further, there are no spectroscopic orbitals farther localized to correct for the fact that the asymptotic behavior of the tail of the wave functions that represent the higher Rydberg states is not well approximated. Thus, the Babushkin gauge that probes the outer part of the wave functions does not produce trustworthy results. The inner parts of the wave functions representing the higher Rydberg states are, however, adequately approximated, and as a result, the Coulomb gauge yields transition rates that are more reliable (see also, Figure 3a,b).

The alternative computational strategy generates correlation orbitals that are more extended, increasing the overlap with the spectroscopic orbitals that make up the higher Rydberg states. In this case, the correlation orbitals are properly localized to ably describe the asymptotic behavior of the outer part of the wave functions representing the higher Rydberg states. That being so, after the final MCDHF and RCI computations in the alternative manner, the reduced matrix elements of the transition operator are practically gauge invariant and the transition rates are also accurately predicted for the transitions between low-lying states and high Rydberg states (see also, Figure 3a,b).

The radial transition integrals (8) and (9) that take part in the computations of the reduced matrix elements of the transition operator have an upper integration bound that goes to infinity. In (8) and (9), $P(r)$ and $P'(r)$ are the radial parts of the spectroscopic and correlation orbitals that are included in the computations. If we express the transition integrals as a function of the upper integration bound R, we get

$$\int_0^R P(r) r P'(r)\, dr \tag{11}$$

and

$$\int_0^R P(r) \frac{d}{dr} P'(r)\, dr, \tag{12}$$

respectively. We can keep $R = \infty$ for the spectroscopic orbitals so that they extend to their full values and only introduce a cut-off value for R in the transition integrals involving correlation orbitals. In this manner, the effect on the transition rate values, from correlation orbitals gradually localized farther from the origin, can be studied. In connection with Figure 3a, the effect that the shape of the correlation orbitals has on the computation of transition rates is, in Figure 4a, examined for the $2p\,^2P^\circ_{1/2} - 8s\,^2S_{1/2}$ transition in C IV. In Figure 4a, the transition rates are computed by employing the alternative computational strategy and both Babushkin and Coulomb gauges are displayed. One observes that the two gauges are affected differently by the outer parts of the correlation orbitals.

In Figure 4a, the transition rate computed in the Coulomb gauge is mainly influenced by the correlation orbitals that are localized close to the origin and in the vicinity of the atomic core. Correlation orbitals occupying regions with $\sqrt{R} > 1$ have an insignificant effect on the Coulomb gauge. This explains the fact that the conventional computational strategy, which generates more contracted orbitals, still predicts with accuracy, the transition rates in the Coulomb gauge. Oppositely, the Babushkin gauge is hugely affected by the correlation orbitals occupying the region between $\sqrt{R} = 5$ and $\sqrt{R} = 6$. Looking at Figure 4b, the $8s$ radial orbital, which extends far out from the $1s^2$ atomic core, begins its asymptotic decay at about $\sqrt{R} = 5$ and dies out at $\sqrt{R} \approx 6$. Only when we have orbitals extending into this region, the asymptotic behavior of the wave function representing the $8s\,^2S_{1/2}$ state is well described, and thus, the Babushkin gauge will yield accurate transition rates. As previously seen, the conventional computational strategy fails to do so.

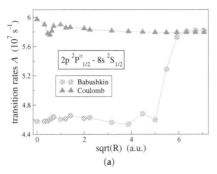
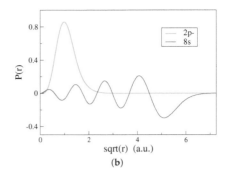

Figure 4. (**a**) The transition rates A in the Babushkin and the Coulomb gauges for the $2p\,^2P^\circ_{1/2} - 8s\,^2S_{1/2}$ transition in C IV, as a function of the square root of the upper integration bound R in the radial transition integrals (11) and (12) involving correlation orbitals. The radial transition integrals involving spectroscopic orbitals extend to their full values, so that $R = 0$ corresponds to transition rates computed from wave functions exclusively built from spectroscopic orbitals. The wave functions are produced by the alternative computational strategy. (**b**) The spectroscopic $2p$ and $8s$ radial orbitals in C IV as a function of $\sqrt{r}$. The two orbitals occupy different regions in space and their overlap is minor. The $8s$ orbital extends far out from the atomic core.

A similar study was performed for a transition between two high Rydberg states. In Figure 5a, the effect of the shape of the correlation orbitals on the computed transition rates is examined for the $7p\,^2P^\circ_{1/2} - 8s\,^2S_{1/2}$ transition in C IV. As seen in Figure 5a, correlation has nearly the same effect on both gauges. Although the $7p$ and $8s$ orbitals extend far out from the atomic core (see Figure 5b), correlation orbitals occupying the large R region remain unimportant in the Babushkin gauge. For transitions between states close in energy, the line strengths take large values and the change in the transition rates due to correlation is very small. For this reason, the conventional computational strategy also yields accurate transition rates for transitions between high Rydberg states.

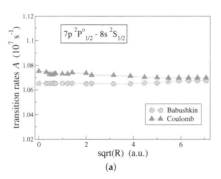
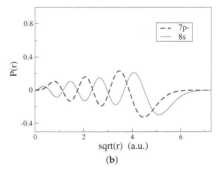

Figure 5. (**a**) Same as Figure 4a, but for the $7p\,^2P^\circ_{1/2} - 8s\,^2S_{1/2}$ transition in C IV. (**b**) The spectroscopic $7p$ and $8s$ radial orbitals in C IV as functions of $\sqrt{r}$. Both orbitals occupy nearly the same regions in space, overlapping to a great extent.

To clarify the fact that the asymptotic behavior of the wave function at large distances is (or is not) well approximated, depending on the alternative (or the conventional) optimization strategy, Brillouin's theorem [33,34] can be put forward to emphasize the importance of the variational content of the wave functions. When being interested into the description of Rydberg states $1s^2nl\,^2L\,(L = l)$ in the single-configuration non-relativistic HF approximation, node counting of the valence radial function, $nc = n - l - 1$, provides a simple and efficient way to select the desired state in the self-consistent procedure [34]. Each separately optimized state implicitly contains all single-electron excitations $nl \rightarrow n'l$ of both lower and upper parts of the spectrum, including the continuum $1s^2\epsilon l$, with the associated interesting property

$$\langle \Phi^{\text{HF}}(1s^2nl\ ^2L)|\mathcal{H}|\Phi(1s^2n'l\ ^2L)\rangle = \langle \Phi^{\text{HF}}(1s^2nl\ ^2L)|\mathcal{H}|\Phi(1s^2\epsilon l\ ^2L)\rangle = 0, \quad \forall(n',\epsilon) \quad (13)$$

where $\mathcal{H}$ is the scalar non-relativistic Hamiltonian that is used in the energy functional to derive the HF equations. The annihilation property of the (M_L, M_S)-independent interaction matrix elements between the reference HF CSF built with the optimized HF orbitals, i.e., Φ^{HF}, and all single-electron excitation CSFs Φ, defines Brillouin's theorem and explains the reasonable accuracy of the HF approximation through the richness of its variational content. The above discussion can be extended to the relativistic framework by considering $n\kappa \rightarrow (n',\epsilon)\kappa$ single-electron excitations and taking for $\mathcal{H}$ the relativistic Hamiltonian that is used for deriving the DHF equations [17].

In the present work, it is hard to define the variational content of the MCDHF approach due to the complexity of the energy functional, but one should keep in mind that the optimization strategy is based (i) on a layer-by-layer approach in which only the last layer is variational while the previous ones are kept frozen, and (ii) on the use of the EOL method targeting, simultaneously, a large number of states for a spectrum calculation. The resulting lack of variational freedom for the individual states can (partially) be counterbalanced by the inclusion of enough interacting states in the Hamiltonian matrix. Going back to the single-configuration approximation case mentioned above, any member of a Rydberg series can be described through configuration interaction involving Brillouin one-electron excitations with a resulting CI-expansion strictly equivalent to the single approximation HF wave function if the basis of single-excitation CSFs is large and rich enough. This equivalence has been exploited to solve convergence problems encountered in the MCHF study of Rydberg series in strontium [35] or to demonstrate the correspondence between different orbital optimization schemes for describing the discrete-continuum interactions in complex systems [36]. In the context of our work, one illustrates the inadequacy of the orbitals obtained in the conventional approach that are used to compensate the lack of variational freedom in the representation of the high-lying Rydberg members. On the contrary, the alternative strategy proposed produces orbitals that have a better localization for describing the single-electron excitations, which would have been implicitly included with a fully optimized MCDHF wave function targeting a single Rydberg state.

6. Summary and Conclusions

The computations of transition data in the systems of lithium-like and beryllium-like carbon are examples of spectrum calculations that involve Rydberg series. In this work, we showed that, independently of the optimization scheme of the radial orbitals, transition parameters corresponding to the lower part of the spectrum are computed with high accuracy. As astronomical spectroscopy raises the demand on atomic data, highly accurate transition parameters are, however, also required for transitions that involve high Rydberg states. We demonstrated how this can be achieved by paying special attention to the optimization scheme of the radial orbitals with respect to correlation. Finally, we showed that the Babushkin gauge should not, by default, be considered as the preferred gauge, and that, in the computations of Rydberg series, it might be required that the transition rates in the Coulomb gauge are used as a reference for the interpretation of astrophysical observations.

Author Contributions: All authors contributed jointly to conceptualization, investigation, validation, and writing—reviewing and editing. A.P. and P.J. cooperatively took part in the methodology and visualization. A.P., P.J., and J.E. conducted the formal analysis. A.P. performed the writing—original draft preparation and data curation.

Funding: This research was funded by the Swedish Research Council (VR) under contract 2015-04842 and 2016-04185. A.P. was financially supported by the Royal Physiographic Society of Lund (https://www.fysiografen.se/en/) to attend the ASOS 2019 conference in Shanghai. M.G. acknowledges support from the FWO-FNRS

Excellence of Science Programme (grant number EOS-O022818F). S.S. is a FRIA grantee of the Fonds de la Recherche Scientifique—FNRS.

Acknowledgments: M.G., G.G., and P.J., were research associates at Vanderbilt University in 1981, 1994–97, and 1995–97, respectively. They are grateful to Charlotte for her endless training and encouragement in many atomic physics research projects over the years, for which her deep knowledge in computational and mathematical physics was crucial. Her outstanding contribution to theoretical atomic physics has been a boundless source of inspiration for all of us.

Conflicts of Interest: The authors declare no conflict of interest.

References

1. Käufl, H.-U.; Ballester, P.; Biereichel, P.; Delabre, B.; Donaldson, R.; Dorn, R.; Fedrigo, E.; Finger, G.; Fischer, G.; Franza, F.; et al. CRIRES: A high-resolution infrared spectrograph for ESO's VLT. *Ground-Based Instrum. Astron.* **2004**, *5492*, 1218–1227.

2. Young, E.T.; Becklin, E.; Marcum, P.M.; Roellig, T.L.; Buizer, J.M.D.; Herter, T.L.; Güsten, R.; Dunham, E.W.; Temi, P.; Andersson, B.-G.; et al. Early Science with SOFIA, the Stratospheric Observatory for Infrared Astronomy. *Astrophys. J. Lett.* **2012**, *749*, L17. [CrossRef]

3. Dorn, R.J.; Anglada-Escude, G.; Baade, D.; Bristow, P.; Follert, R.; Gojak, D.; Grunhut, J.; Hatzes, A.; Heiter, U.; Hilker, M.; et al. CRIRES+: Exploring the Cold Universe at High Spectral Resolution. *Messenger* **2014**, *156*, 7–11.

4. Atomic Databases from the Opacity Project and the Iron Project. Available online: http://cdsweb.u-strasbg.fr/topbase/testop/home.html (accessed on 13 November 2019).

5. Galvez, F.J.; Buendia, E.; Sarsa, A. $1s^2 2p^3$ and $1s^2 2s^2 3l$, $l = s, p, d$, excited states of boron isoelectronic series from explicitly correlated wave functions. *J. Chem. Phys.* **2005**, *123*, 034302.

6. Papoulia, A.; Ekman, J.; Jönsson, P. Extended transition rates and lifetimes in Al I and Al II from systematic multiconfiguration calculations. *Astron. Astrophys.* **2019**, *621*, A16. [CrossRef]

7. Crossley, R.J.S. The Calculation of Atomic Transition Probabilities. *Adv. At. Mol. Phys.* **1969**, *5*, 237–296. [CrossRef]

8. Starace, A.F. Length and Velocity Formulas in Approximate Oscillator-Strength Calculations. *Phys. Rev. A* **1971**, *3*, 1242.

9. Starace, A.F. Comment on "Length and Velocity Formulas in Approximate Oscillator-Strength Calculations". *Phys. Rev. A* **1973**, *8*, 1141. [CrossRef]

10. Grant, I.P. Gauge invariance and relativistic radiative transitions. *J. Phys. B At. Mol. Opt. Phys.* **1974**, *7*, 1458. [CrossRef]

11. Grant, I.P.; Starace, A.F. Gauge invariance and radiative transition probabilities. *J. Phys. B At. Mol. Opt. Phys.* **1975**, *8*, 1999. [CrossRef]

12. Crossley, R.J.S. Fifteen Years On—The Calculation of Atomic Transition Probabilities Revisited. *Phys. Scr.* **1984**, *T8*, 117–128. [CrossRef]

13. Hibbert, A. Oscillator strengths of transitions in the beryllium sequence. *J. Phys. B At. Mol. Opt. Phys.* **1974**, *7*, 1417. [CrossRef]

14. Cowan, R.D. Electric dipole radiation. In *The Theory of Atomic Structure and Spectra*; University of California Press: London, UK, 1981; pp. 400–402. [CrossRef]

15. Pehlivan Rhodin, A.; Hartman, H.; Nilsson, H.; Jönsson, P. Experimental and theoretical oscillator strengths of Mg I for accurate abundance analysis. *Astron. Astrophys.* **2017**, *598*, A102.

16. Grant, I.P. *Relativistic Quantum Theory of Atoms and Molecules*; Springer: New York, NY, USA, 2007. [CrossRef]

17. Froese Fischer, C.; Godefroid, M.; Brage, T.; Jönsson, P.; Gaigalas, G. Advanced multiconfiguration methods for complex atoms: I. Energies and wave functions. *J. Phys. B At. Mol. Opt. Phys.* **2016**, *49*, 182004.

18. Verdebout, S.; Jönsson, P.; Gaigalas G.; Godefroid, M.; Froese Fischer, C. Exploring biorthonormal transformations of pair-correlation functions in atomic structure variational calculations. *J. Phys. B At. Mol. Opt. Phys.* **2010**, *43*, 074017. [CrossRef]

19. Dyall, K.G.; Grant, I.P.; Johnson, C.T.; Parpia, F.A.; Plummer, E.P. GRASP: A general-purpose relativistic atomic structure program. *Comput. Phys. Commun.* **1989**, *55*, 425–456. [CrossRef]

20. Gaigalas, G.; Rudzikas, Z.; Froese Fischer, C. An efficient approach for spin-angular integrations in atomic structure calculations. *J. Phys. B At. Mol. Opt. Phys.* **1997**, *30*, 3747–3771. [CrossRef]

21. Gaigalas, G.; Fritzsche, S.; Grant, I.P. Program to calculate pure angular momentum coefficients in jj-coupling. *Comput. Phys. Commun.* **2001**, *139*, 263–278. [CrossRef]
22. Froese Fischer, C.; Gaigalas, G.; Jönsson, P.; Bieroń, J. GRASP2018-A Fortran 95 version of the General Relativistic Atomic Structure Package. *Comput. Phys. Commun.* **2019**, *237*, 184–187. [CrossRef]
23. Froese Fischer, C.; Godefroid, M.R. Programs for computing *LS* and *LSJ* transitions from MCHF wave functions. *Comput. Phys. Commun.* **1991**, *64*, 501–519. [CrossRef]
24. Froese Fischer, C. Evaluating the accuracy of theoretical transition data. *Phys. Scr.* **2009**, *T134*, 014019. [CrossRef]
25. Ekman, J.; Godefroid, M.; Hartman, H. Validation and Implementation of Uncertainty Estimates of Calculated Transition Rates. *Atoms* **2014**, *2*, 215–224. [CrossRef]
26. Froese Fischer, C.; Brage, T.; Jönsson, P. Structure of ψ_1. In *Computational Atomic Structure—An MCHF Approach*; CRC Press: Boca Raton, FL, USA, 1997; pp. 70–71. [CrossRef]
27. Olsen, J.; Roos, B.O.; Jørgensen, P.; Jensen, H.J.A. Determinant based configurationinteraction algorithms for complete and restricted configuration interaction spaces. *J. Chem. Phys.* **1988**, *89*, 2185.
28. Sturesson, L.; Jönsson, P.; Froese Fischer, C. JJGEN: A flexible program for generating lists of jj-coupled configuration state functions. *Comput. Phys. Commun.* **2007**, *177*, 539–550. [CrossRef]
29. Godefroid, M.; Jönsson, P.; Froese Fischer, C. Atomic structure variational calculations in spectroscopy. *Phys. Scr.* **1998**, *T78*, 33–46. [CrossRef]
30. Froese Fischer, C.; Jönsson, P.; Godefroid, M. Some two-electron properties of sodium. *Phys. Rev. A* **1998**, *57*, 1753. [CrossRef]
31. Froese Fischer, C.; Brage, T.; Jönsson, P. Term dependence. In *Computational Atomic Structure—An MCHF Approach*; CRC Press: Boca Raton, FL, USA, 1997; pp. 55–56. [CrossRef]
32. Kramida, A.; Ralchenko, Y.U.; Reader, J.; NIST ASD Team. *NIST Atomic Spectra Database (ver. 5.6.1)*; National Institute of Standards and Technology: Gaithersburg, MD, USA, 2018. Available online: https://physics.nist.gov/asd (accesssed on 24 September 2019).
33. Godefroid, M.; Liévin, J.; Metz, J.-Y. Brillouin's theorem for complex atomic configurations. *J. Phys. B At. Mol. Opt. Phys.* **1987**, *20*, 3283–3296.
34. Froese Fischer, C. *The Hartree-Fock Method for Atoms. A Numerical Approach*; John Wiley and Sons: New York, NY, USA, 1977. [CrossRef]
35. Vaeck, N.; Godefroid, M.; Hansen, J.E. Multiconfiguration Hartree-Fock calculations for singlet terms in neutral strontium. *Phys. Rev. A* **1988**, *38*, 2830–2845.
36. Cowan, R.D.; Hansen, J.E. Discrete-continuum interactions in Cl I and S I. *JOSA* **1981**, *71*, 60. [CrossRef]

 *atoms*

MDPI

Article

Charlotte Froese Fischer—Her Work and Her Impact

Alan Hibbert

Centre for Theoretical Atomic, Molecular and Optical Physics, School of Mathematics & Physics, Queen's University, Belfast BT7 1NN, UK; a.hibbert@qub.ac.uk

Received: 21 October 2019; Accepted: 11 December 2019; Published: 17 December 2019

Abstract: Charlotte Froese Fischer has been at the forefront of research in atomic structure theory for over 60 years. She has developed many of the methods currently used by researchers and has written associated computer programs which have been published and hence made accessible to the research community. Throughout her career, she has consistently encouraged and mentored young scientists, enabling them to embark on independent careers of their own. This article provides an overview of the methods and codes she has developed, some large-scale calculations she has undertaken, and some insight into the impact she has had on young scientists, and the leadership she continues to show as she reaches her 90th birthday.

Keywords: multiconfigurational methods; MCHF; MCDHF; GRASP; non-orthogonal orbitals; B-splines; E1 transitions; forbidden transitions; hyperfine structure; isotope shifts

1. Foundations

Charlotte Froese was born in the Ukraine, but soon after her birth, her family moved, via Germany, to Canada. She grew up near to Vancouver, and undertook her undergraduate and masters studies, in Mathematics along with Chemistry and some Physics, at the University of British Columbia (UBC) in Vancouver [1]. From there, she moved to the University of Cambridge, England, to undertake studies for her PhD under the supervision of Douglas Hartree. This interaction provided Charlotte (and therefore us) with a link to the very beginnings of atomic structure calculations. These had begun with the foundational paper of Schrödinger (1926) [2], which also dealt with the very simple case of hydrogenic ions. This was extended by Hylleraas (1928) [3] to the case of the ground state of helium, in which he incorporated the interelectronic distance r_{12} explicitly in a form of the wave function which also contained variational parameters, determined by minimising the total energy. The use of interelectronic coordinates, while possible for very simple atomic systems, did not lend itself to extension to the calculations for many-electron atoms and ions. Instead, Hartree (1927, 1928) [4,5] proposed a method which was capable of application to any number of electrons in an atom or ion. In this, the N-electron wave function is represented by a product of N one-electron functions, or *orbitals*, with the motion of an individual electron determined by a single orbital.

$$\tilde{\psi}(1, 2, .., N) = u_1(1)u_2(2)...u_N(N) \tag{1}$$

He used physical arguments to demonstrate that the orbital functions $u_i(i)$ should satisfy the non-linear equations

$$\left(-\frac{1}{2}\nabla_i^2 - \frac{Z}{r_i}\right) u_i(r_i) + \sum_{j\neq i} u_i(r_i) \int \frac{|u_j(r_j)|^2}{r_{ij}} d\tau_j = \varepsilon_i u_i(r_i) \tag{2}$$

He chose to take a spherical average of the final term of the potential so that the angular dependence of the orbital functions took the form of single spherical harmonics, Y_l^m, and also the radial parts of the orbital functions were assumed to be independent of m. The number of unknown functions was thereby substantially reduced. Nevertheless, the non-linearity of the equations meant that they had to be solved iteratively, with the aim of achieving self-consistency.

Hartree's simple product form of the wave function did not satisfy the anti-symmetry requirement for the wave function. Fock (1930) [6] therefore wrote the wave function as an *anti-symmetrised* product of orbitals, represented by a determinant:

$$\Psi(LS) = \frac{1}{\sqrt{N!}} \begin{vmatrix} \phi_1(1) & \cdots & \phi_N(1) \\ \vdots & \ddots & \vdots \\ \phi_1(N) & \cdots & \phi_N(N) \end{vmatrix} \tag{3}$$

in which the one-electron functions (spin-orbitals) now incorporated a function describing the spin of the electron:

$$\phi_i(r_i, m_{si}) = u_i(r_i)[\alpha(i) \text{ or } \beta(i)] \tag{4}$$

with $m_s = \pm\frac{1}{2}$ respectively.

This led to more elaborate equations—the Hartree-Fock equations—particularly because of the additional term in the potential, representing the possible exchange of the indistinguishable electrons:

$$\left(-\frac{1}{2}\nabla_i^2 - \frac{Z}{r_i}\right)u_i(r_i) + \sum_{j \neq i}\left[u_i(r_i)\int \frac{|u_j(r_j)|^2}{r_{ij}}d\tau_j - \delta(m_{s_i}, m_{s_j})u_j(r_i)\int \frac{u_j^*(r_j)u_i(r_j)}{r_{ij}}d\tau_j\right] = \varepsilon_i u_i(r_i) \tag{5}$$

Comparing the potential terms in Equations (2) and (5), that in Equation (2) has the same form as the first of the two terms in Equation (5). It is referred to as the *direct term*, for it comprises the direct interaction between electron i and the field arising from all the other electrons. Equation (5) contains an additional term, arising from the antisymmetric nature of the wave function. Compared with the first term, it can be seen that the function u_i has been interchanged with one of the u_j, so that it appears inside the integral in the second term. Because of this exchanging of the placing of these orbital functions, this second term is referred to as the *exchange term*. Its physical interpretation is that it models the indistinguishability of electrons, a feature missing from Hartree's formalism. In the case of Hartree's Equation (2), the potential is *local*, in that the equations can be expressed as

$$\left(-\frac{1}{2}\nabla_i^2 - \frac{Z}{r_i}\right)u_i(r_i) + V(r_i)u_i(r_i) = \varepsilon_i u_i(r_i)$$

where V, representing the potential, does not involve u_i. By contrast, such a multiplicative potential cannot be formed for Equation (5), since u_i appears inside the integral in the second term. Therefore the potential in this case is *non-local*.

In both cases, the equations are non-linear in the orbital functions, so must be solved iteratively. Typically, at any iteration stage, the opening forms of the orbital functions are used to evaluate the integrals, and so the integro-differential equations become linear differential equations. For (2), each equation then involves just a single orbital function, whereas for (5), the exchange term means that the equations are coupled, involving all the orbital functions. In both cases, the orbitals comprising the solution at each stage of iteration become the input orbitals for the next stage, although in practice the orbitals obtained at earlier stages might also be used in the formation of the input orbitals, in order to speed up the convergence towards self-consistency.

Hartree set to work on solving, again self-consistently, these more elaborate equations, and although Fock's formalism did not require it, he chose to represent the angular dependence of the space part of the orbital functions again by single spherical harmonics, and the radial functions as being independent of m_l and m_s. The Hartree-Fock equations then became equations for the radial parts of the orbital functions.

For the next 25 years or so Hartree, alone (e.g., Hartree 1934 [7]) or with his father (Hartree and Hartree 1938) [8] and with other co-workers (e.g., Hartree, Hartree and Swirles 1939 [9]), determined wave functions for a variety of atoms and ions, with the radial functions being in tabular numerical form. It was a laborious undertaking, demanding careful book-keeping of the set of numerical functions, even though Hartree was later able to enlist some of the first computers that became available, and indeed which he was instrumental in designing [10].

2. Hartree-Fock Calculations

It was into this scientific environment that Charlotte came in the mid-1950s. Her first paper with Hartree concerned the solution of the Hartree-Fock (HF) equations for Ne IV and Ne V, (Froese and Hartree 1957 [11]). It was important to understand in detail the steps which had to be taken, in a systematic manner, for self-consistency to be achieved, in the solution of the HF equations. Charlotte's mathematical background equipped her well for the tasks of studying the equations themselves and for programming the EDSAC computer in Cambridge. Sadly, it was to be her only paper with Hartree, for he died early in 1958, shortly after his definitive book on the subject (Hartree 1957 [12]) was published. His untimely death was a significant loss to the atomic structure community and of course to Charlotte herself. But, picking up his mantle, she proceeded to publish further papers during the same year ([13,14]), in which she completed the numerical solutions of the HF equations for a number of ions, including some of astrophysical importance. Over the next few years, further studies of the HF equations followed, but it is interesting to note her focus on aspects of the mathematical or numerical solution of the equations, for example on applications to high Z-values ([15–17]). By then, Charlotte had returned to UBC in Vancouver, where a new computer had just been installed—the first at UBC. She was then able to follow up the approaches she had developed with Hartree during her time in Cambridge, by investigating the numerical solution of the HF equations with emphasis on how electronic computers could be used most effectively, as well as considering the ways in which accuracy could be assured [18]. She was a pioneer in the use of electronic computers in atomic structure calculations, and was becoming a world leader in the field—a position she still holds!

The 1960s proved to be very significant for Charlotte. In 1964, she was awarded an Alfred P. Sloan Fellowship, a prestigious award given to those early-career researchers who show exceptional promise. She was the first woman to be given that award and accolade. Clearly, the foresight of the awarders has been fully vindicated!

Further HF calculations followed in those years, for a range of atoms and ions, of wave functions and energies, and other atomic properties such as oscillator strengths and hyperfine structure. In addition, issues affecting the rate of convergence of the iterative process in the solution of the HF equations as well as other mathematical aspects of the HF equations were studied and resolved.

Above all, she met and married Patrick Fischer, and in 1968 they moved across Canada to the computer science department of the University of Waterloo, Ontario. Since then, she has published under the name of Charlotte Froese Fischer.

3. Extensions of HF

This detailed study of the mathematics of the solution of the HF equations, as well as her development of an algorithmic approach, stood her in good stead for taking forward some of the many ways of extending the HF method in order to achieve greater agreement with experimental results, in energy differences, but also in the calculation of other properties such as oscillator strengths, hyperfine structure or isotope shifts.

Hartree's original method as well as the HF method can be expressed in terms of a variational approach to calculating energies (Slater 1930 [19]). Consequently, HF energies are reasonably accurate, at least for isolated and low-lying states, since the errors in the energies are of second-order for first-order changes in the orbitals. However, in the case of other operators (for example the dipole operator encountered in the calculation of oscillator strengths), the errors in their matrix elements are of first order.

A powerful and now widely used means of improving on HF is through the use of wave functions which, in *LS* angular momentum coupling, are of configuration interaction format:

$$\Psi(LS) = \sum_{i=1}^{M} a_i \, \Phi_i(\alpha_i LS) \tag{6}$$

where α_i represents the coupling of the angular momenta of the orbitals in each of the configuration state functions (CSFs) Φ_i, and typically Φ_1 is chosen to be the HF wave function. For any choice of the form of the CSFs, the optimal values of the mixing coefficients a_i are eigenvector components of the Hamiltonian matrix whose typical element is $< \Phi_i | H | \Phi_j >$, with the corresponding set of eigenvalues providing the calculated energy values. Froese (1964) [20] demonstrated the insufficiency of the HF process in her study of multiplet transitions in Fe XV and Fe XVI. The Fe XVI ion, is essentially hydrogenic with a single electron outside completely closed shells, and the HF method leads to reasonably accurate oscillator strengths. By contrast, for some transitions in Fe XV, HF is not adequate. Froese found that considerable improvement could be achieved by the inclusion of configuration interaction, particularly for $^1P^o - {}^1D$ transitions. Even if the lower state 3s3p $^1P^o$ is represented by just the HF configuration, there are two possible upper levels—3s3d and 3p^2, and each state needs to be represented by a linear combination of the two ^1D configurations, and both configurations have large mixing coefficients in each state. The radial functions of the orbitals were generated in HF calculations on different states, and then used in a configuration interaction, or more precisely superposition of configurations, calculation. Hartree, Hartree and Swirles (1939) [9] had incorporated configuration interaction much earlier, albeit with the simpler system of oxygen ions. Froese incorporated the mixing coefficients in the Hartree-Fock equations, thus providing an early example of the multi-configuration Hartree-Fock (MCHF) calculation.

In the MCHF process, the variational method is used to generate the MCHF equations, using a trial wave function in the form of equation (i6), and by setting to zero the first-order change in the energy expression $< \Psi | H | \Psi >$ subject to the orthonormality of the orbital functions. The MCHF equations are similar to the HF equations in that they are the equations which determine the orbital functions, but additionally incorporate the optimisation of the mixing coefficients, a_i of Equation (6). The MCHF method includes the additional process of the diagonalisation of the Hamiltonian matrix, thus providing the optimal values of the mixing coefficients for the current forms of the orbital functions.

The key difference between MCHF and the superposition of configurations method is that in MCHF, the orbital functions are obtained as the solutions of the MCHF equations, whereas generally the term *configuration interaction* implies that the orbital functions are pre-determined separately, though with the same the diagonalisation of the Hamiltonian matrix to provide the CI mixing coefficients. For example, in an MCHF calculation of a light element such as Be I, the MCHF orbitals as well as the CI mixing coefficients would be determined directly by the two-stage process of solving the MCHF equations (to a pre-determined level of consistency) followed by the diagonalisation of the Hamiltonian matrix. The CI or superposition of configurations process might first fix those orbital functions which are 'occupied' in the HF approximation, with further orbitals, added from a variety of sources, to be used in the other CSFs. A bringing together of these somewhat different processes is sometimes undertaken for larger atoms, when, for example, the orbitals for some of the subshells might be fixed from simpler calculations, such as those of their HF forms or from an MCHF calculation using just the CSFs with substantial mixing coefficients, while the additional orbitals introduced for the other CSFs are treated as unknowns, to be determined by solving the MCHF equations. As a consequence, generally the MCHF wave functions are more accurate than CI wave functions, because of use of the variational principle in setting up the MCHF equations, though the difference can be fairly small.

A major application of the use of MCHF or CI wave functions is in the calculation of oscillator strengths of transitions for atoms or ions. The oscillator strength, or f-value, for electric dipole transitions is a dimensionless quantity, and therefore is usually evaluated in atomic units, and for an N-electron atom or ion can be expressed, in *velocity* form, as

$$f_v = \frac{2}{3}\frac{1}{g_i \Delta E} \left| \left\langle \psi_j \left| \sum_{k=1}^{N} \nabla_k \right| \psi_i \right\rangle \right|^2 \tag{7}$$

where ΔE, given in atomic units, is the transition energy (the energy difference between the two states) and g_i is the g-value of the energetically lower state, which means that in *LS* coupling $g_i = (2L_i + 1)(2S_i + 1)$. The emission *transition rate* A^{ji}, sometime called the *transition probability*, is related to the absorption oscillator strength f^{ij} by, in the case of electric dipole transitions,

$$f^{ij} = 1.4997 \times 10^{-16} \lambda^2 \frac{g_j}{g_i} A^{ji}$$

with A expressed in units of s^{-1}, and λ (in Å) is the wavelength of the transition. The equivalent *length* form of the oscillator strength is

$$f_l = \frac{2}{3}\frac{1}{g_i}\Delta E \left| \left\langle \psi_j \left| \sum_{k=1}^{N} r_k \right| \psi_i \right\rangle \right|^2 \tag{8}$$

Oscillator strengths appear in astrophysical modelling, for example in the determination of element abundances in stellar atmospheres, in the combination $\log(gf)$ so oscillator strengths are frequently published as gf-values, with g as the g-value of the lower state/level of the transition.

When the wave functions of the two states involved are exact eigenfunctions of the Hamiltonian, so that the Hamiltonian commutes with each r_k, the length and velocity forms give the same result. For HF, the potential is non-local, so these conditions are not satisfied. Hence in most cases, the calculated length and velocity forms do not agree for HF wave functions. However, when a very simple local model potential is used, with the two wave functions each being exact eigenfunctions of this simple Hamiltonian, then length and velocity forms do agree, but the common value is not necessarily correct. So, while it is necessary for length and velocity forms to agree, their doing so is not a guarantee of accuracy. Rather, accuracy is achieved by studying the convergence of the two forms as the numbers of CSFs in expansions (6) of the CI or MCHF wave functions are increased. It is sometimes argued that the length value is the more reliable of the two, and so only the length is calculated or provided. It is indeed often the case that the length value is the more stable as wave functions are extended, as the velocity form is more affected by the degree of electron correlation included in the calculation. But to omit the velocity value is to remove one measure of accuracy. If the two forms converge at least closely to a common value, one can have confidence in the accuracy of that value. It is this convergence process that Charlotte has endeavoured to achieve through her calculations.

4. Some Illustrative Examples

Configuration interaction calculations were undertaken for the some challenging transitions in Al I (Froese 1965) [21] and for Si II (Froese and Underhill 1966) [22]. This work was extended [23] to include a consideration of the very different 3d radial functions when optimised on the two ^{2}D states separately, and the effect of imposing conditions which ensure the orthogonality of the two ^{2}D states. The results for Si II are shown in Table 1.

Table 1. CI calculations for Si II.

CI Coefficients for ^{2}D States of Si II			
Configurations	$3s3p^2$	$3s^23d$	$3s^23d$ (orthog)
$3s3p^2$	0.7908	-0.5629	-0.5263
$3s^23d$	0.5994	0.8016	0.8251
$3p^2(^1S)3d$	0.1146	0.1921	0.1977
$3p^2(^1D)3d$		0.0118	
$3p^2(^3P)3d$		-0.0603	-0.0563
$3s3d^2$	0.0467		
gf-values from the $3s^23p$ $^2P^o$ ground state			
Froese Fischer (1968) [23]	0.103		6.22
Froese Fischer (1981) [24]	0.006		6.83
Hibbert et al. (1992) [25]	0.011		6.69

In the first part of the table, it is clear that the two main ^{2}D configurations have strong components in each state. In an extension of [22], Froese Fischer [23] found state the mean radius of the optimal MCHF 3d function when optimised on the $3s3p^2$ state was 3.215, whereas when optimised on the $3s^23d$ state, the mean radius of 3d was 5.412. When the orthogonality constraint is imposed, some changes in the mixing coefficients occur, but the strong CI mixing in both states is generally maintained. The strong CI mixing has a pronounced effect on the oscillator strengths (shown in Table 1 as gf values, with $g = (2L + 1)(2S + 1)$ for the state with lower energy = 6). The oscillator strength between the ground state and the $3s^23d$ state is enhanced by the CI mixing, whereas that in the transition to the $3s3p^2$ state, strong CI cancellation occurs, resulting in an abnormally small gf value. In a much more extensive calculation of these transitions, involving many configurations, including those representing core polarisation [24], and the use of non-orthogonal orbitals to circumvent the problem of different 3d functions being optimal for the two ^{2}D states, the cancellation is stronger still. That very small oscillator strength agrees well with an independent calculation of my own [25]. This transition is of astrophysical significance in the study of the abundance of Si in the interstellar medium, Shull et al. (1981) [26]. They determined, from observations, a gf value of around 0.033, not quite in agreement with the most recent calculations but, given the extent of the cancellation, it can be seen that calculated values and results derived from observation are fairly close.

One important feature of Charlotte's MCHF calculations is the demonstration of convergence of results as the CI expansions of the wave functions are increased in size. While there is a variational principle which ensures that a longer expansion leads to a monotonic lowering of energy for a particular wave function, there is no such guarantee of monotonic improvement either for energy *differences* or for oscillator strengths. One way of assessing the accuracy of results is to see the way in which these quantities change as wave function expansions are extended in a systematic manner. Such a systematic analysis is provided by Tong et al. (1995) [27] for low-lying quartet transitions in neutral nitrogen. The basic way of providing a systematic analysis is first by constructing an active set of correlation orbitals usually characterised by specifying the maximum n-value and allowing all possible orbitals up to that maximum value. Ideally, the active space should consist of all possible CSFs which can be formed using the active set of orbitals, but this would be a prohibitively large calculation. Instead, Tong et al. [27] undertook a development of a sequence of increasingly complex models. in which the active set is systematically increased.

The most challenging transition is $2s^22p^3$ $^4S^o$ − $2s2p^4$ ^4P, because of quite severe CI cancellation in the transition integral.

Some results are shown in Table 2, demonstrating ways in which the convergence of results as the wave functions are increased in complexity can be studied. One of the features of this transition is that the $2s2p^4$ ^4P state interacts very strongly with two other ^4P states—$2s^22p^23s$ and $2s^22p^23d$—so that all three states must be treated in an equivalent manner. In the first part of the table, the increase in complexity is achieved by extending the type of correlation effects included. Thus, model 1 includes single and double replacements of $n = 2$ orbitals by an increasing number (up to $n = 6$) of correlation orbitals (the active space) in the reference set comprising $2s^2 2p^3$ and $2s2p^4$. Correlation effects are therefore not included in an equivalent way in the three ^{4}P states. This defect is corrected in model 2, which includes configurations $2s^2 2p^2 3l$ in the reference set. Model 3 allows for CSFs with only one orbital from the {2s,2p} set. In this analysis, a major improvement in the oscillator strength is achieved through model 2. It is worth noting that even in model 1, the length and velocity forms of the oscillator strength are in fairly good agreement (they differ only by about 3%), but the common value is clearly incorrect. In the second half of the table, we see how Tong et al. [27] undertook a systematic analysis of oscillator strength and transition energy values as the size of the active set is increased, within model 3. At each stage, defined by the value of the largest n-value of the orbitals included in that stage, all orbitals are computed using the MCHF program. Only when the $n = 5$ orbitals does any sense of convergence appear. The change between $n = 6$ and $n = 7$, the latter denoted by model 3+ in the first part of the table, is quite small. This analysis is characteristic of Charlotte Froese Fischer's approach towards achieving confidence in the accuracy of the results of her calculations.

Table 2. Oscillator strengths for the $2s^2 2p^3$ ^{4}S$^{\rm o}$ $-$ $2s2p^4$ ^{4}P transition in N I [27].

Model Number	Configuration Complexes †	ΔE	f_l	f_v
1	$\{2\}^3$ $\{2,3,\ldots,6\}^2$	87,271	0.3513	0.3628
2	$\{2\}^2$ $\{2,3\}^1$ $\{2,3,\ldots,6\}^2$	88,524	0.0533	0.0563
3	$\{2\}^1$ $\{2,3\}^2$ $\{2,3,\ldots,6\}^2$	88,375	0.0667	0.0693
3+	$\{2\}^1$ $\{2,3\}^2$ $\{2,3,\ldots,7\}^2$	88,356	0.0658	0.0687
Within model 3	$\{2\}^1$ $\{2,3\}^2$ $\{2,3\}^2$	89,760	0.3163	0.4062
	$\{2\}^1$ $\{2,3\}^2$ $\{2,3,\ldots,4\}^2$	89,324	0.1108	0.1171
	$\{2\}^1$ $\{2,3\}^2$ $\{2,3,\ldots,5\}^2$	88,446	0.0701	0.0717
	$\{2\}^1$ $\{2,3\}^2$ $\{2,3,\ldots,6\}^2$	88,375	0.0667	0.0693
	$\{2\}^1$ $\{2,3\}^2$ $\{2,3,\ldots,7\}^2$	88,356	0.0658	0.0687
Exp. [28] Average over J		88,132		

$\dagger$: The notation indicates the orbital occupancy of the 5 outer electrons; for example $\{2\}^2$ $\{2,3\}^1$ $\{2,3,\ldots,6\}^2$ means two orbitals with $n = 2$, plus one from $n= 2$ or 3; plus two with any n value from 2 through to 6; additionally, all CSFs contain $1s^2$.

5. Computer Programs for Atomic Structure

The publication of her HF code [18] was but the first example of her adoption of the principle that if computer codes were of wide applicability, they should be made available to other users and not simply retained for personal use. There was clearly a need to create a computer program for the MCHF method, similar to that of the HF method [18], in which the processes were again automated once the CSFs and the initial estimates for the radial functions of the orbitals were selected, and which could readily be extended to any number of CSFs, or to any atomic system. During the 1960s, Charlotte developed such a code. The energy functional was based on Equation (6), from which the MCHF equations were derived using the variational principle. As with the HF process, these equations were coupled integro-differential equations, but now the mixing coefficients a_i were included. Again, the solution comprised an iterative process. At each iteration, the integrals in the MCHF equations were calculated from the previous iteration of the radial functions, and also the mixing coefficients were taken from the previous iteration; the resulting (coupled) differential equations were then solved to give a new set of radial functions, and with these new functions, the new mixing coefficients were obtained by diagonalising the Hamiltonian matrix, although in early versions of the MCHF code only one eigenvalue/eigenvector was allowed, with the eigenvector components computed iteratively. Thus the orbital radial functions were obtained directly from these MCHF equations, rather than from separate HF solutions, followed by a superposition of configurations process, as had been used in earlier calculations. Hence the MCHF orbitals themselves incorporated the effects of configuration interaction.

In 1969 a new journal, *Computer Physics Communications (CPC)*, was launched, with my colleague in Belfast, Phil Burke, as Principal Editor and Charlotte as one of the subject editors for atomic structure. This journal provided the vehicle for the dissemination of the MCHF code (Froese Fischer 1969 [29]), updated a little later with the name MCHF72 (Froese Fischer 1972 [30]), and again later still (Froese Fischer 1978 [31]). This journal provided assurance for both authors and users of the published codes. Submissions to the journal required a description of the methods underlying the codes, as well as details of how the codes were constructed—for example, the subroutines, procedures, modules from which the code was built and how they linked together. In addition, sample input data had to be provided, as well as the corresponding output and a statement of which type of computer had been used in obtaining the output, and the size of the program. The codes and the papers describing them were then rigorously refereed before acceptance for publication. Users could then compare the output from the test data when they ran the code on their own machine against the output submitted with the code. They could then be reasonably well assured that the code of interest was what they required, and that it worked as expected. The authors had much more to do in preparing the paper and code for publication than is customary for a standard research paper, but having done so, they were much less likely to be asked for a copy of the code, or questioned about its reliability than if they had distributed the code privately. The underlying assumption was (and is) that the authors' codes become widely available for others to use.

There are benefits to authors as well as safeguards for users. For example, their codes become known and used by a wider community, so that due recognition of the authors' work is accorded. Unexpected or unforeseen 'bugs' might be eliminated at the journal's checking stage. Requests for copies of the codes can be directed to the journal.

One additional benefit I have found, both as a user and authors, is that users and authors might and sometimes do work together on further code developments. It was this aspect that led me to begin working actively with Charlotte Froese Fischer. The MCHF code aims to solve the integro-differential equations for the radial functions of the orbitals. However, the Hamiltonian matrix elements in the energy functional can be expressed as a weighted sum of radial integrals. The integrals over the angular and spin coordinates can be achieved exactly and these data form part of the input to the code. My own code, WEIGHTS, also published in the first volume of *CPC* (Hibbert 1970 [32]), together with a follow-up the following year (Hibbert 1971 [33]), provided the weighting coefficients for the two-electron part of the Hamiltonian, while a later code (Hibbert 1974 [34]) provided the same data for the one-electron operators. This led to a strong collaboration between Charlotte and myself, which has continued from time to time ever since.

In the early years, it was not self-evident that the new journal would be a success. Some sceptics doubted whether authors would be willing to spend the time necessary for a thorough description of their codes, or to subject them to the rigorous refereeing scrutiny that was required. Others wondered how willing users would be to familiarise themselves with the detailed working of the codes. Their fears were unfounded, as is evidenced by the fact that as the 50th anniversary of the journal arrives, the volume count is almost at 250. This success is due in no small measure to the enthusiasm and determination of Phil Burke, Charlotte Froese Fischer and others who were well established researchers in their fields and whose involvement gave great credence to the value of the journal and to the quality of the papers and programs it would be publishing: their names alone gave the new journal considerable credibility. Over the next 50 years or so, Charlotte was to publish around 40 programs or procedures in *CPC*, and throughout that time, she maintained full support for the principles on which the journal was developed, and of course her codes were very robust and reliable.

6. Extensions and Enhancements of MCHF—Non-Relativistic Treatment

We consider here the extensions to the MCHF process, but still in the context of calculations in *LS* angular momentum coupling. We will report on a relativistic approach in the following section.

6.1. Non-Orthogonal Orbitals

In using the MCHF process, it is customary to express all the atomic states in terms of a common set of orbitals which are orthogonal to each other. But this is a restriction, which for elements with only a small number of electrons outside closed shells can sometimes be overcome with additional configuration interaction (essentially by extending the set of orbitals), but in other cases even very extensive additional CI would be required and possibly the restriction cannot fully be overcome even then. It is therefore necessary to consider the possibility of using orbitals which, for the same *l*-value, are not orthogonal to each other (orbitals with different *l*-values are of course mutually orthogonal).

Some exemplars of where non-orthogonal orbitals are useful include:

1. He: $1s^2$ and $1s2p$—radially, the 1s function of the 1s2p state is close to being hydrogenic whereas the 1s function for the ground state resembles a screened hydrogenic function.
2. Be: $[1s^2]2s^2$ ^1S, $2s2p$ $^3P^o$ and $^1P^o$—the 2s functions differ somewhat from state to state, but the more significant feature is that the mean radii of the 2p functions in the two excited states differ by around a factor of two.
3. Al-sequence: we have already noted in Table 1 that the optimal 3d function in the two lowest ^2D states is very state dependent. A more appropriate CI expansion would have configurations of the form:

$$^2D: 3s_1^2 3d;\ 3p_1^2(^1S)3d,\ 3p_2^2(^1D,^3P)3d_1,\ 3s_2 3p_3^2$$

where the same *nl* orbital but with different subscripts need not be mutually orthogonal.
4. The 3d orbital in open d-shells, for example in the iron group elements, can be very term-dependent even for an individual ion.

The lifting of the restriction that the same orbital set, comprising mutually orthogonal functions, be used for all states can lead to substantial improvements in the accuracy of the results, and/or much shorter CI expansions to achieve comparable effects. For example, even in the light element neutral nitrogen, Robinson and Hibbert (1997) [35] found that just a few configurations could achieve for quartet transitions an accuracy as good as, and in some cases much better—comparing length and velocity values—than could be obtained with a much larger calculation using orthogonal orbitals (Hibbert et al. 1985) [36]. The difficulty was obtaining agreement between the length and velocity forms, and although the calculations of Robinson and Hibbert were not definitive, they did achieve comparable agreement with Tong et al. (1994) [27], who found that it was necessary to use of some thousands of CSFs. Some results are shown in Table 2.

To be able to undertake CI or more specifically MCHF calculations in the framework of non-orthogonal orbitals, I renewed my collaboration with Charlotte as we worked out how to modify the codes which undertake the angular and spin integrals, in order to incorporate the possibility of using non-orthogonal orbitals. This resulted in work which was published in *CPC* (Hibbert et al. 1988) [37], work which also began my collaboration with Michel Godefroid (Brussels). The extent of the non-orthogonality was limited: essentially we considered various pair correlations in any atomic state and allowed non-orthogonality of the orbitals in different pairs.

In order to study transitions, it was also necessary to allow for non-orthogonality between the orbitals in the two states or levels in the transition. In particular, the two wave functions used in the calculation of the oscillator strength might be calculated separately, using different orbitals in each case. This situation was allowed for by the use of a bi-orthonormal transformation of the orbital functions by which the methods applicable to orthogonal orbitals can be used, before calculating the transition matrix elements. The general theory of this process was introduced by Olsen et al. (1995) [38], in which the application to the MCHF codes was developed by Godefroid.

Its power was amply demonstrated by a study of the $2s^22p\ ^2P^\circ - 2s2p^2\ ^2D$ transition in boron. An example, given by Olsen et al. (1995) [38], of the success of the use of non-orthogonal orbitals, when combined with the bi-orthonormal transformation of the orbitals is shown in Table 3 for the $2s^22p\ ^2P^\circ - 2s2p^2\ ^2D$ transition in neutral boron.

Table 3. Oscillator strengths for the $2s^22p\ ^2P^\circ - 2s2p^2\ ^2D$ transition in B I [38].

Active Set: max n	$\Delta E(\text{cm}^{-1})$	gf_l	gf_v
3	53,197	0.6876	0.8156
4	48,720	0.2456	0.2606
5	48,440	0.2625	0.2695
6	48,125	0.2891	0.2866
7	48,051	0.2928	0.2900
7 E	47,847	0.2916	0.2912
Other results			
Method		gf_l	gf_v
MCHF [39]		0.243	0.274
Expt: LIF [40]		0.283 ± 0.020	

E: Using a weighted average (over J) of the individual levels given in [28].

Table 3 exemplifies the process of making systematic improvements to the calculations. Specifically, the concept of active space of orbitals is used, consisting of all possible orbitals with an increasing value of n. The reference set of configurations consisted of $(1s^22s^22p, 1s^22p^3;\ ^2P^\circ)$, $(1s^22s2p^2, 1s^22s^23d;\ ^2D)$. The CSF set included all which are obtained by single and double replacements of the orbitals occupied in the reference set by any from the active set. The $n = 3$ results allow only for a single 3d orbital, whereas the optimal 3d from the $1s^22s^23d$ differs substantially from the optimal 3d of $1s^22s2p^2$, and that difference is not included at that stage of the calculation. This is substantially rectified by $n = 4$. The trend from $n = 4$ to $n = 7$ shows a systematic improvement in the transition energy ΔE and the agreement between the length and velocity forms of the gf-values. The final results, labelled 7^E, are obtained from the $n = 7$ results by using the experimental rather than calculated transition energy. We also give in Table 3 the MCHF results using orthogonal orbitals [39] and the experimental result of O'Brian and Lawler [40] who used the laser-induced fluorescence method. The improvement obtained by using non-orthogonal orbitals can be clearly seen, and the final calculation of gf lies well inside the experimental error bars. (Incidentally, the MCHF result of [39] is an example of the velocity form being closer than the length form to the experimental or converged calculated value of the oscillator strength, a counter-example to the view that the length value is the better of the two).

6.2. Use of B-Splines

The MCHF method focuses on the best possible way of introducing short-range electron correlation (assuming *LS* coupling) into the solution of the Schrödinger equation. It works well when the active electrons (those in the outermost shells) have rather similar mean radii, so that electron exchange is a significant effect. Equally, it works well for excited, but not too highly excited, states. But for studying Rydberg series in atoms and ions, where the outer electron has a much larger mean radius than all the other electrons, then electron exchange has a lower probability and representation of highly excited states is not so easy to accomplish with the customary MCHF approach. Indeed, Hartree's original method, which ignores electron exchange, becomes an improving approximation. One of the difficulties of the use of orthogonal orbitals in MCHF calculations is that an MCHF orbital nl with a high n-value contains oscillations in the radial function arising from the requirement of orthogonality to those with lower n, which the use of non-orthogonal orbitals would not introduce. To overcome this difficulty, Froese Fischer and co-workers (e.g., Brage and Froese Fischer 1994 [41]) have modified the MCHF method by using B-splines to represent the orbitals, and particularly those of the outermost electrons. The number of B-splines required can be adjusted to encompass the radial range even of the outermost orbital. Each B-spline is a relatively localised function, being non-zero over only part of the radial range covered by the orbitals, so each orbital function must be represented by a sum of appropriate B-splines. While the B-splines are mutually orthogonal, the orbitals represented by these sums are not necessarily orthogonal, so a non-orthogonal approach is necessary. This flexibility can lead to more accurate representations of the wave functions. This use of B-splines is part of a vast field of study well described by Bachau et al. (2001) [42]. An example of the use of B-splines in atomic structure is given by Brage and Froese Fischer (1994) [41]. They studied several Rydberg series in neutral calcium. They allowed for some limited non-orthogonality of the outer orbitals and added the effect of core polarisation by means of a model potential. A small selection of their results is included in Table 4.

Table 4. Binding energies (cm^{-1}) of 4snp $^1\text{P}^\text{o}$ states in Ca II [41].

Label	Exp [28]	MCHF+BS	MCHF
4s4p	25,654	25,472	24,689
4s5p	12,574	12,684	12,160
4s6p	7627	7638	7060
3d4p	5372	5269	4609
4s7p	3881	3799	3247
4s8p	2826	2786	2405
4s9p	2122	2102	1853
. . .			
4s 22p	271	271	

It can be seen that the inclusion of the effects of core polarisation, together with the use of a B-spline representation of the outer orbitals, (MCHF+BS), gives a substantially improved agreement with experiment for the binding energies, compared with the conventional MCHF approach. In particular, at the upper end of the Rydberg series, the agreement is excellent.

There are distinct similarities between this approach and that adopted in R-matrix calculations [43,44], where the outer (free) electron is customarily represented by a linear combination of basis set of *continuum* orbitals. Most R-matrix calculations to date have required the continuum orbitals to be orthogonal to the orbitals describing the bound orbitals of the N-electron core, but Zatsarinny and Froese Fischer [45] have undertaken an R-matrix calculation of the photoionisation of Li using B-splines and non-orthogonal orbitals.

7. Inclusion of Relativity

The MCHF method is essentially non-relativistic. But in order to study both allowed and forbidden transitions among levels, it was necessary for Charlotte to incorporate fine-structure effects into the calculation. Two approaches were available: a multi-configurational formalism based on the Dirac equation rather than Schrödinger's equation, or an approximation to this approach in which the non-relativistic Schrödinger Hamiltonian is augmented by the operators of the Breit-Pauli Hamiltonian. For light atoms and ions, the latter process is adequate for the accuracy required in many applications. For heavy elements as well as for highly ionised systems with fewer electrons, the fully relativistic approach based on the Dirac equation is normally required.

7.1. Breit-Pauli Calculations

Since many of the applications being considered were for the lighter elements, Froese Fischer chose the Breit-Pauli approach. Her calculations followed the following pattern.

- The radial functions were optimised in an *LS* MCHF calculation.

- The angular and spin integrals of the relativistic operators were evaluated using the Racah algebra analysis given by Glass and Hibbert (1978) [46] and are input to the MCHF+BP code.
- Then the full Breit-Pauli Hamiltonian matrix was diagonalised to give the *LSJ* wave functions. These wave functions take the form

$$\Psi(J) \;=\; \sum_{i=1}^{M} a_i\, \Phi_i(\alpha_i L_i S_i J) \tag{9}$$

so that CSFs with different L_i and S_i can be combined to a common total *J*.

Initially, Charlotte studied fine structure separations of low-lying term energies and of the forbidden magnetic dipole and electric quadrupole transitions between them. For example, she calculated the splitting between the $^2P^o_{0.5}$ and $^2P^o_{1.5}$ of boron-like ions [47]. Some results are shown in Table 5 and compared with the fully relativistic treatment.

Table 5. Term splitting (cm^{-1}) of the ground state in B-like ions.

Method	B	N^{2+}	Ne^{5+}	Si^{9+}	Fe^{21+}
MCHF+BP [47]	15.0	170.0	1292	6961	119,175
MCDHF [a]	15.7	172.4	1298	6968	118,177
Exp [b]	15.3	174.5	1307	6990	118,255

[a] Huang et al. (1982) [48]; [b] NIST [28].

It can be seen that the Breit-Pauli approximation gives results which are very close to the fully relativistic results of [48] near the neutral end of the sequence but, as expected, diverge as *Z* increases, whereas the fully relativistic calculations are consistently close to the experimental values.

Similar calculations followed for the ground terms of other fairly light elements: C([49]), N([50]) and O([51]), and these papers also included the rates of forbidden transitions between the ground term levels. Around the same time, electric dipole transitions were studied, using the same MCHF+Breit-Pauli formalism [52], and this work permitted the study of intercombination lines. Of particular interest was the calculation of the $2s^2\ ^1S_0 - 2s2p\ ^3P^o_1$ line in C III [53], as displayed in Table 6.

Table 6. MCHF+BP transition rates of the $2s^2\ ^1S_0 - 2s2p\ ^3P^o_1$ line in C III [53].

Degree of Correlation	Active Space	ΔE (cm^{-1})	$^1P^o_1 - {}^3P^o_1$	$^3P^o_2 - {}^3P^o_0$	A (s^{-1})
Val [a]	$n = 3$	52,746	51,592	76.79	89.3
	$n = 6$	52,733	50,684	77.08	95.6
+CP [b]	$n = 3$	52,640	50,567	78.06	97.6
	$n = 6$	52,520	50,098	80.18	105.7
+CC [c]	$n = 3$	52,362	50,948	77.33	91.9
	$n = 6$	52,343	50,230	79.53	103.1
CIV3 [54]		52,369	50,325	78.9	103.8
MCDHF [55]	$n = 8$	52,384	50,098	79.86	102.72
Experiment	[28]	52,391	49,961	80.05	
	[56]				121 ± 7
	[57]				102.94 ± 0.14

[a] Valence correlation only; [b] a + core polarisation; [c] b + partial core correlation.

As in other calculations undertaken by Charlotte, a systematic development of the results can be seen, both in the sense of growing complexity in the type of correlation included and, with each type, the variation in results as the active space is enlarged. The final result differs from experiment by about 20%, but is in complete agreement with another, independent calculation [54]. A later calculation [55] gave a transition rate of 102.72, which when extrapolated to take into account the slight inaccuracies in the *ab initio* energy separations resulted in a recommended value of 102.87, lying within the very narrow error bars of the recent heavy-ion storage ring experiment [57]. This calculation was updated by Froese Fischer and Gaigalas [58] to yield a transition rate of 103.0 s^{-1}, with an estimated uncertainty of 0.4 s^{-1}.

Many further calculations followed, for different ions, and this work culminated in the extensive tabulations of energy levels and electric dipole oscillator strengths by Froese Fischer and Tachiev (2004) [59] for the first row elements and their ions, and by Froese Fischer et al. (2006) [60] for the second row elements and their ions. These important compendia provide a set of accurate atomic data for transitions between a substantial number of levels of these elements. The data are characteristic of Charlotte's work: they are undertaken in a consistent manner, and consider all the main correlation and relativistic effects appropriate for these elements. In the discussion in the papers, there is a strong attention to detail and where possible, comparison is made with experiment, especially the energy levels. They demonstrate why her work is considered to be first-rate and reliable for other researchers, including astrophysical modellers, to use with confidence.

7.2. Other Atomic Properties

Once MCHF+BP wave functions are available, it is possible to determine atomic properties other than transition rates. For example, Charlotte undertook calculations of isotope shifts and hyperfine interactions [61], photoionisation [62], and autoionisation [63].

7.3. Further MCHF-Based Computer Packages

By the 1980s, *CPC* had begun to include papers on computational methods, as well as continuing to publish computer codes and their descriptions. Accordingly, in order to pull together in one place several of the developments in MCHF and BP procedures which had taken place during the 1980s, Charlotte requested that almost the whole of one issue of *CPC* was devoted to these developments. The details are summarised in Table 7.

Table 7. MCHF and Associated Codes.

Authors	Short Title	Type of Paper
Froese Fischer [64]	The MCHF atomic structure package	Methods
Froese Fischer [65]	MCHF support libraries and utilities	Package
Froese Fischer, Liu [66]	Configuration-state lists	Package
Hibbert, Froese Fischer [67]	Angular integrals with non-orthogonal orbitals	Package
Froese Fischer [68]	General MCHF program	Package
Hibbert, Glass, Froese Fischer [69]	Angular integrals for Breit-Pauli Hamiltonian	Package
Froese Fischer [70]	General CI program	Package
Froese Fischer, Godefroid, Hibbert [71]	Angular integrals for transition operators	Package
Froese Fischer, Godefroid [72]	Programs for *LS* and *LSJ* transitions	Package

These codes constitute a comprehensive package of codes allowing users to undertake a wide range of MCHF calculations. Once again, Charlotte was demonstrating her continuing commitment to making codes available to other users, codes which had been thoroughly tested.

8. Fully Relativistic Codes

The use of the Breit-Pauli Hamiltonian works well for lighter ions, but for transitions in heavy ions or for heavy elements, the use of fully relativistic wave functions becomes more accurate. In a multiconfigurational context, Desclaux (1975) [73] published the multiconfiguration Dirac-Fock (MCDF) code, based on a Dirac rather than Schrödinger formalism. This nomenclature of MCDF lacks the acknowledgement of the importance of Hartree's work, and Charlotte preferred the designation MCDHF, the 'H' denoting 'Hartree'. An enhanced version, GRASP, was published by Dyall et al. (1987) [74]. One of the authors (Parpia) spent some time working with Charlotte and as a consequence, Charlotte began to work with Grant's group, culminating in an updated version of GRASP: Parpia et al. (1996) [75,76], with a further update in 2007 [77]. Further enhancements followed and were published in *CPC*, including the calculation of other properties using relativistic wave functions, such as hyperfine structure [78] and isotope shifts [79].

Some of the enhancements involved the use of more efficient methods for the calculation of the angular momentum integrals in GRASP. In this, Charlotte made use of the opportunity to collaborate with the group in Vilnius, whose work she had long admired [80]. The following 20 years saw the publication of many calculations of atomic properties for medium and heavy ions. Often, these were the outcome of an on-going international research team—CompAS—comprising Charlotte and groups based in Lund, Malmö, Vilnius, Krakow, Gdansk and Brussels. The group meets periodically to review progress and plan for future collaborations. I was present for part of their discussions in Lund in 2018 and in Brussels in 2019. I found Charlotte thoroughly involved in those discussions, an amazing degree of engagement at the age of 90.

9. In Summary

Charlotte Froese Fischer has been at the forefront of atomic structure developments for over six decades. In her early years in research, she developed programs for the calculation of atomic properties using electronic computers, which were only just becoming available. In subsequent years, the initial Hartree-Fock methods were extended, characterised by

$$HF \rightarrow MCHF \rightarrow MCHF+BP \rightarrow MCHF+BS \rightarrow Non\text{-}orthogonal \rightarrow MCDHF \rightarrow GRASP.$$

Charlotte has consistently been keen to use the most up-to-date computer architecture available to her, and to develop new numerical methods to exploit such facilities. This work involved the study of the convergence of MCHF and MCDHF iterations, and the efficient calculation of the angular momentum integrals (which are one of the most time-consuming parts of the calculations). Additionally, efficient methods and codes for the diagonalisation of huge matrices were developed. For large-scale MCHF or MCDHF calculations, with many thousands of CSFs included in the CI expansion of the wave functions, the Hamiltonian matrices are very large, but only a relatively small number of eigenvalues/eigenvectors, those which are the lowest in energy, are needed for subsequent calculations. The iterative approach, initially developed by Davidson [81], was further developed and programmed for this purpose [82].

Charlotte was part of the original team of scientists which set up the journal *Computer Physics Communications* in 1969. The publication of her codes in this journal has amply demonstrated her commitment to the ethos of the journal, not least in ensuring that others can make direct use of her work in undertaking their own calculations. That commitment was one of the reasons why the journal was able to develop in its early years, because other scientists could see the value of publishing their programs in that new medium. The ongoing success of the journal is a tribute to the commitment of Charlotte and the other editors, and of course to the long-term vision of its first Principal Editor Phil Burke.

As well as her original research papers, Charlotte has also published a number of textbooks, suitable for graduate students, which explain the methods she has developed and provide a detailed explanation of how calculations are undertaken. Amongst these are a discussion of the general Hartree-Fock method [83], and more recently an explanation of the computational approach to solving the MCHF equations and the calculation of atomic properties [84]. These books facilitate other scientists, including those starting out in the field, in developing an expertise in undertaking atomic structure calculations for themselves.

Charlotte is the holder of a number of awards recognising her international reputation as a world leader in the field of atomic physics. The Alfred P. Sloan Fellowship, awarded in 1964, was an early recognition of her promise. She was elected Fellow of the American Physical Society in 1991 for her contribution to the discovery of the calcium negative ion and for her extensive and innovative researches. Her research standing was also recognised overseas. In 1995, she was elected Fellow of the Royal Physiographic Society in Lund, Sweden. In retirement, her research continues, and in 2004 she was elected a foreign member of the Lithuanian Academy of Sciences. Then as recently as 2015 she was awarded an honorary doctorate by the University of Malmö, while in 2016 she was elected a Fellow of the Royal Society of Canada, especially appropriate as she returned to Vancouver.

But it is not just for her scientific contributions to the field of atomic structure, immense though they are, that Charlotte should be applauded. She has also had a large influence on the careers of many scientists, particularly on younger colleagues. As well as demonstrating her openness in making her computer codes available to the entire community, and in providing support for users, she has been keen to work with young scientists and to be a mentor to them, through the provision of encouragement as well as guidance on the standards of research expected in the field. And she has continued to collaborate with those who were young scientists and who have been able to develop their careers and their activity in the field. As Charlotte reaches her 90th birthday, her enthusiasm and commitment continue unabated.

Charlotte Froese Fischer is a remarkable woman, and I consider it a privilege to be able to count her as a colleague and friend.

Funding: This research received no grant funding as such, but I wish to acknowledge the generous support of the organisers of the ASOS2019 conference to enable me to attend and present the talk on which this paper is based.

Acknowledgments: Charlotte Froese Fischer kindly read an earlier version of the manuscript of this paper and made a number of helpful suggestions.

Conflicts of Interest: The author declares no conflict of interest.

References

1. Froese Fischer, C. Reminiscences at the end of the century. *Mol. Phys.* **2000**, *98*, 1043–1050. [CrossRef]
2. Schrödinger, E. Quantisierung als eigenwertproblem. *Ann. Phys.* **1926**, *79*, 361–376. [CrossRef]
3. Hylleraas, E.A Über den Grundzustand des Heliumatoms. *Z. Phys.* **1928**, *48*, 469–494. [CrossRef]
4. Hartree, D.R. The Wave Mechanics of an Atom with a Non-Coulomb Central Field. Part I. Theory and methods. *Proc. Camb. Philos. Soc.* **1927**, *24*, 89–110. [CrossRef]
5. Hartree, D.R. The Wave Mechanics of an Atom with a Non-Coulomb Central Field. Part II. Some Results and Discussion. *Proc. Camb. Philos. Soc.* **1928**, *24*, 111–132. [CrossRef]
6. Fock, V.A. Näherungsmethode zur Lösung des quantenmechanischen Mehrkörperproblems. *Z. Phys.* **1930**, *61*, 126–148. [CrossRef]
7. Hartree, D.R. Approximate wave functions and atomic field for mercury. *Phys. Rev.* **1934** *46*, 738–743. [CrossRef]
8. Hartree, D.R.; Hartree, W. Wave functions for negative ions of sodium and potassium. *Proc. Camb. Philos. Soc.* **1938**, *34*, 550–558. [CrossRef]
9. Hartree, D.R.; Hartree, W.; Swirles, B. Self-consistent field, including exchange and superposition of configurations, with some results for oxygen. *Philos. Trans. R. Soc.* **1939**, *A238*, 229–247. [CrossRef]
10. Hartree, D.R. The differential analyser. *Nature* **1935**, *135*, 940–943. [CrossRef]
11. Froese, C.; Hartree, D.R. Wave functions for the normal states of Ne^{3+} and Ne^{4+}. *Proc. Camb. Philos. Soc.* **1957**, *53*, 663–668. [CrossRef]
12. Hartree. D.R. *The Calculation of Atomic Structures*; John Wiley and Sons: New York, NY, USA, 1957; pp. xiii+181.
13. Froese, C. The self-consistent field with exchange for the ground state and first excited state of Fe^{13+}. *Mon. Not. R. Astron. Soc.* **1957**, *117*, 615–621. [CrossRef]
14. Froese, C. The self-consistent field with exchange for some 10 and 12 electron systems. *Proc. Camb. Philos. Soc.* **1957**, *53*, 206–213. [CrossRef]

15. Froese, C. The limiting behaviour of atomic wave functions for large atomic number. *Proc. R. Soc. Ser. A* **1957**, *239*, 311–319.

16. Froese, C. The limiting behaviour of atomic wave functions for large atomic number. II. *Proc. R. Soc. Ser. A* **1958**, *244*, 390–397.

17. Froese, C. The limiting behavious of atomic wave functions for large atomic number. III. *Proc. R. Soc. Ser. A* **1959**, *251*, 534–535.

18. Froese, C. Numerical solution of the Hartree-Fock equations. *Can. J. Phys.* **1963**, *41*, 1895–1910. [CrossRef]

19. Slater, J.C. Note on Hartree's method. *Phys. Rev.* **1930**, *35*, 210. [CrossRef]

20. Froese, C. Some multiplet strengths for transitions in Fe XVI and Fe XV. *Astrophys. J.* **1964**, *140*, A1489–A1494. [CrossRef]

21. Froese, C. Oscillator strengths for the $3s^23p\ ^2P - 3s3p^2\ ^2S$ transition in Al I. *Astrophys. J.* **1965**, *141*, 1557–1559. [CrossRef]

22. Froese, C.; Underhill, A.B. gf-values for lines of the Si II spectrum. *Astrophys. J.* **1966**, *146*, 301–303. [CrossRef]

23. Froese Fischer, C. Superposition of configuration results for Si II. *Astrophys. J.* **1968**, *151*, 759–764. [CrossRef]

24. Froese Fischer, C. Lifetimes of low-lying $^2P^o$, ^2D, and $^2F^o$ states of the Al I isoelectronic sequence. *Phys. Scr.* **1981**, *23*, 38–44. [CrossRef]

25. Hibbert, A.; Ojha, P.C.; Stafford, R.P. Allowed transitions in Si II. *J. Phys. B At. Mol. Opt. Phys.* **1992**, *25*, 4153–4162. [CrossRef]

26. Shull, J.M.; Snow, T.P.; York, D.G. Observationally determined Silicon II oscillator strengths. *Astrophys. J.* **1981**, *246*, 549–553. [CrossRef]

27. Tong, M.; Froese Fischer, C.; Sturesson, L. Systematic transition probability studies for neutral nitrogen. *J. Phys. B At. Mol. Opt. Phys.* **1994**, *27*, 4819–4828. [CrossRef]

28. Kramida, A.; Ralchenko, Y.; Reader, J.; NIST ASD Team. *NIST Atomic Spectra Database (ver. 5.3)*; National Institute of Standards and Technology: Gaithersburg, MD, USA, 2018. Available online: http://physics.nist.gov/asd (accessed on 26 July 2019).

29. Froese Fischer, C. A Multiconfiguration Hartree-Fock program. *Comput. Phys. Commun.* **1969**, *1*, 151–166. [CrossRef]

30. Froese Fischer, C. A Multiconfiguration Hartree-Fock program with improved stability. *Comput. Phys. Commun.* **1972**, *4*, 107–116. [CrossRef]

31. Froese Fischer, C. A general multi-configuration Hartree-Fock program. *Comput. Phys. Commun.* **1978**, *14*, 145–153. [CrossRef]

32. Hibbert, A. A general program for calculating angular momentum integrals in atomic structure. *Comput. Phys. Commun.* **1970**, *1*, 359–377. [CrossRef]

33. Hibbert, A. A new version of a general program to calculate angular momentum integrals. *Comput. Phys. Commun.* **1971**, *2*, 180–190. [CrossRef]

34. Hibbert, A. Adaptation of a program to calculate angular momentum integrals: Inclusion of the one-electron part of the Hamiltonian. *Comput. Phys. Commun.* **1974**, *7*, 318–326. [CrossRef]

35. Robinson, D.J.R.; Hibbert, A. Quartet transitions in neutral nitrogen. *J. Phys. B At. Mol. Opt. Phys.* **1997**, *30*, 4813–4825. [CrossRef]

36. Hibbert, A.; Dufton, P.L.; Keenan, F.P. Oscillator strengths for transitions in N I and the interstellar abundance of nitrogen. *Mon. Not. R. Astron. Soc.* **1985**, *213*, 721–734. [CrossRef]

37. Hibbert, A.; Froese Fischer, C.; Godefroid, M.R. Non-orthogonal orbitals in MCHF or configuration interaction wave functions. *Comput. Phys. Commun.* **1988**, *51* 285–293. [CrossRef]

38. Olsen, J.; Godefroid, M.R.; Jönsson, P.; Malmqvist, P.Å.; Froese Fischer, C. Transition probability calculations for atoms using nonorthogonal orbitals. *Phys. Rev. E* **1995**, *52*, 4499–4508. [CrossRef]

39. Carlsson, J.; Jönsson, P.; Sturesson, L.; Froese Fischer, C. Lifetimes and transition probabilities of the boron atom calculated with the active space multiconfiguration Hartree-Fock method. *Phys. Rev. A* **1994**, *49*, 3426–3431. [CrossRef] [PubMed]

40. O'Brian, T.R.; Lawler, J. Radiative lifetimes in B I using ultraviolet and vacuum-ultraviolet laser-induced fluorescence. *Astron. Astrophys.* **1992**, *255*, 420–426.

41. Brage, T.; Froese Fischer, C. Spline-Galerkin calculations for Rydberg series calculations of calcium. *Phys. Scr.* **1994**, *49*, 651–660. [CrossRef]

42. Bachau, H.; Cormier, E.; Decleva, P.; Hansen, J.E.; Martin, F. Application of B-splines in atomic and molecular physics. *Rep. Prog. Phys.* **2001**, *64*, 1815–1942. [CrossRef]

43. Burke, P.G.; Hibbert, A.; Robb, W.D. Electron scattering by complex atoms. *J. Phys. B At. Mol. Phys.* **1971**, *4*, 153–161. [CrossRef]

44. Berrington, K.A.; Burke, P.G.; Le Dourneuf, M.; Robb, W.D.; Taylor, K.T.; Lan, V.K. A new version of the general program to calculate atomic continuum processes using the R-matrix method. *Comput. Phys. Commun.* **1978**, *14*, 367–412. [CrossRef]

45. Zatsarinny, O.; Froese Fischer, C. The use of B-splines and non-orthogonal orbitals in R-matrix calculations: Application to Li Photoionization. *J. Phys. B At. Mol. Opt. Phys.* **2000**, *33*, 313–341. [CrossRef]

46. Glass, R.; Hibbert, A. Relativistic effects in many-electron atoms. *Comput. Phys. Commun.* **1978**, *16*, 19–34. [CrossRef]

47. Froese Fischer, C. Multiconfiguration Hartree-Fock Breit-Pauli results for $^2P_{1/2}$-$^2P_{3/2}$ transitions in the boron sequence. *J. Phys. B At. Mol. Opt. Phys.* **1983**, *16*, 157–165. [CrossRef]

48. Huang, K.-N.; Kim, Y.-K.; Cheng, K.T.; Desclaux, J.P. Correlation and relativistic effects in spin-orbit splitting. *Phys. Rev. Lett.* **1982**, *48*, 1245–1248. [CrossRef]

49. Froese Fischer, C.; Saha, H.P., Multiconfiguration Hartree-Fock results with Breit-Pauli corrections for transitions in the carbon sequence. *Phys. Scr.* **1985**, *32*, 181–194. [CrossRef]

50. Godefroid, M.; Froese Fischer, C. MCHF-BP fine structure splittings and transition rates for the ground configuration in the nitrogen sequence. *J. Phys. B At. Mol. Opt. Phys.* **1984**, *17*, 681–692. [CrossRef]

51. Froese Fischer, C.; Saha, H.P. Multiconfiguration Hartree-Fock results with Breit-Pauli corrections for forbidden transitions in the $2p^4$ configuration. *Phys. Rev. A* **1983**, *28*, 3169–3178. [CrossRef]

52. Froese Fischer, C.; Saha, H.P. MCHF+BP results for electric dipole transitions in the oxygen isoelectronic sequence. *J. Phys. B At. Mol. Opt. Phys.* **1984**, *17*, 943–952. [CrossRef]

53. Froese Fischer, C. Allowed transitions and intercombination lines in C III and C II. *Phys. Scr.* **1994**, *49*, 323–330. [CrossRef]

54. Fleming, J.; Hibbert, A.; Stafford, R.P. The 1909Å Intercombination Line in C III. *Phys. Scr.* **1994**, *49*, 316–322. [CrossRef]

55. Jönsson, P.; Froese Fischer, C. Multiconfiguration Dirac-Fock calculations of the $2s^2$ $^1S_0$ – $2s2p$ $^3P_1$ intercombination transition in C III. *Phys. Rev. A* **1998**, *57*, 4967–4970. [CrossRef]

56. Kwong, V.H.S.; Fang, Z.; Gibbons, T.T.; Parkinson, W.H.; Smith, P.L. Measurement of the transition probability of the C III 190.9 nanometer intersystem line. *Astrophys. J.* **1993**, *411*, 431–437. [CrossRef]

57. Doerfert, J.; Träbert, E.; Wolf, A.; Schwalm, D.; Uwira, O. Precision measurement of the electric dipole intercombination rate in C^{2+}. *Phys. Rev. Lett.* **1997**, *78*, 4355–4358. [CrossRef]

58. Froese Fischer, C.; Gaigalas, G. Note on the $2s^2$ $^1S_0$ – $2s2p$ $^3P_1$ intercombination line of B II and C III. *Phys. Scr.* **1997**, *56*, 436–438. [CrossRef]

59. Froese Fischer, C.; Tachiev, G. Breit–Pauli energy levels, lifetimes, and transition probabilities for the beryllium-like to neon-like sequences. *Atom. Data Nucl. Data Tables* **2004**, *87*, 1–184. [CrossRef]

60. Froese Fischer, C.; Tachiev, G.; Irimia, A. Relativistic energy levels, lifetimes, and transition probabilities for the sodium-like to argon-like sequences. *Atom. Data Nucl. Data Tables* **2006**, *92*, 607–812. [CrossRef]

61. Jönsson, P.; Froese Fischer, C.; Godefroid, M.R. Accurate calculations of transition probabilities, isotope shifts and hyperfine structures for some allowed $2s^22p^n$–$2s2p^{n+1}$ transitions in B I, C II and C I. *J. Phys. B At. Mol. Opt. Phys.* **1996**, *29*, 2393–2412. [CrossRef]

62. Froese Fischer, C.; Saha, H.P. Photoionization of magnesium. *Can. J. Phys.* **1987**, *65*, 772–776. [CrossRef]

63. Froese Fischer, C.; Idrees, M. Autoionization rates for core excited ^5P states of Na^-. *Phys. Scr.* **1989**, *39*, 70–72. [CrossRef]

64. Froese Fischer, C. The MCHF atomic-structure package. *Comput. Phys. Commun.* **1991**, *64*, 369–398. [CrossRef]

65. Froese Fischer, C. MCHF atomic-structure package: Support; libraries and utilities. *Comput. Phys. Commun.* **1991**, *64*, 399–405. [CrossRef]

66. Froese Fischer, C.; Liu, B. A program to generate configuration-state lists. *Comput. Phys. Commun.* **1991**, *64*, 405–416. [CrossRef]

67. Hibbert, A.; Froese Fischer, C. A general program for computing angular integrals of the non-relativistic Hamiltonian with non-orthogonal orbitals. *Comput. Phys. Commun.* **1991**, *64*, 417–430. [CrossRef]

68. Froese Fischer, C. A general multi-configuration Hartree-Fock program. *Comput. Phys. Commun.* **1991**, *64*, 431–454. [CrossRef]
69. Hibbert, A.; Glass, R.; Froese Fischer, C. A general program for computing angular integrals of the Breit-Pauli Hamiltonian. *Comput. Phys. Commun.* **1991**, *64*, 455–472. [CrossRef]
70. Froese Fischer, C. A configuration interaction program. *Comput. Phys. Commun.* **1991**, *64*, 473–485. [CrossRef]
71. Froese Fischer, C.; Godefroid, M.R.; Hibbert, A. A program for performing angular integrations for transition operators. *Comput. Phys. Commun.* **1991**, *64*, 486–500. [CrossRef]
72. Froese Fischer, C.; Godefroid, M.R. Programs for computing *LS* and *LSJ* transitions from MCHF wave functions. *Comput. Phys. Commun.* **1991**, *64*, 501–519. [CrossRef]
73. Desclaux, J.P. A multiconfiguration relativistic Dirac-Fock program. *Comput. Phys. Commun.* **1975**, *9*, 31–45. [CrossRef]
74. Dyall, K.G.; Grant, I.P.; Johnson, C.T.; Parpia, F.A.; Plummer, E.P. GRASP: A general-purpose relativistic atomic structure program. *Comput. Phys. Commun.* **1987**, *55*, 425–456. [CrossRef]
75. Parpia, F.A.; Froese Fischer, C.; Grant, I.P. GRASP92: A package for large-scale relativistic atomic structure calculations. *Comput. Phys. Commun.* **1996**, *94*, 249–271. [CrossRef]
76. Froese Fischer, C.; Gaigalas, G.; Ralchenko, Y. Some corrections to GRASP92. *Comput. Phys. Commun.* **2006**, *175*, 738–744. [CrossRef]
77. Jönsson, P.; He, X.; Froese Fischer, C.; Grant, I.P. The GRASP2K relativistic atomic structure package. *Comput. Phys. Commun.* **2007**, *177*, 597–622. [CrossRef]
78. Jönsson, P.; Parpia, F.A.; Froese Fischer, C. HFS92: A program for relativistic atomic hyperfine structure calculations. *Comput. Phys. Commun.* **1996**, *96*, 301–310. [CrossRef]
79. Jönsson, P.; Froese Fischer, C. SMS92: A program for relativistic isotope calculations. *Comput. Phys. Commun.* **1997**, *100*, 81–92. [CrossRef]
80. Gaigalas, G.; Rudzikas, Z.; Froese Fischer, C. An efficient approach for spin-angular integrations in atomic structure calculations. *J. Phys. B At. Mol. Opt. Phys.* **1997**, *30*, 3747–3771. [CrossRef]
81. Davidson, E.R. The iterative calculation of a few of the lowest eigenvalues and corresponding eigenvectors of large real-symmetric matrices. *J. Comput. Phys.* **1975**, *17*, 87–94. [CrossRef]
82. Stathopoulos, A.; Froese Fischer, C. A Davidson program for finding a few selected extreme eigenpairs of a large, sparse, real, symmetric matrix. *Comput. Phys. Commun.* **1994**, *79*, 268–290. [CrossRef]
83. Froese Fischer, C. *The Hartree-Fock Method for Atoms*; Wiley Interscience: New York, NY, USA, 1977; pp. xi+320.
84. Froese Fischer, C.; Brage, T.; Jönsson, P. *Computational Atomic Structure—An MCHF Approach*; Institute of Physics Publishing: Bristol, UK, 1997; pp. xi+279.

Article

New Energy Levels and Transitions of 5s²5p² (6d+7s) Configurations in Xe IV

Jorge Reyna Almandos [1,*]**, Mónica Raineri** [1]**, Cesar J. B. Pagan** [2] **and Mario Gallardo** [1]

[1] Centro de Investigaciones Opticas (CIOp), CC 3, 1897, Gonnet, 1900 La Plata, Argentina;
    monicar@ciop.unlp.edu.ar (M.R.); mogallardo38@gmail.com (M.G.)
[2] School of Electrical and Computer Engineering, University of Campinas (UNICAMP), 13083-970 Campinas,
    SP, Brazil; pagan@unicamp.br
* Correspondence: jreyna@ciop.unlp.edu.ar

Received: 17 October 2019; Accepted: 11 December 2019; Published: 17 December 2019

Abstract: Three-times ionized xenon Xe IV spectrum in the 1070–6400 Å region was analyzed using a pulsed discharge light source. A set of 163 transitions was classified for the first time, and 36 new energy levels belonging to the 5s²5p²6d and 5s²5p²7s even configurations were determined. The relativistic Hartree–Fock method, including core-polarization effects, were used. In these calculations, the electrostatic parameters were optimized by a least-square procedure in order to improve the adjustment to experimental energy levels. We also present a calculation based on a relativistic multiconfigurational Dirac–Fock approach.

Keywords: atomic databases and related topics; astrophysical and laboratory plasmas: atomic data needs; atomic lifetime and oscillator strength determination

---

1. Introduction

There is great interest in spectroscopy data of Xenon due to their applications in collision physics, astrophysics, and laser physics. Various atomic parameters, such as energy levels, oscillator strengths, transition probabilities, and radiative lifetimes, have many important astrophysical applications. Transition probabilities are needed for calculating the energy transport through the star in model atmospheres [1] and for direct analysis of stellar chemical compositions [2]. Xenon was observed in chemically peculiar stars [3] and planetary nebulae [4]. The spectrum analysis of planetary nebula NGC7027 by Péquignot and Baluteau [5] has stimulated the calculation of transition probabilities for some forbidden lines of astrophysical interest [6]. The Xe VI and Xe VII lines were observed in the ultraviolet spectrum of the hot DO-type white dwarf RE 0503-289 [7,8]. In particular, the Xe IV spectrum was detected in the spectrum of NGC 7027 together with a variety of ionic species, providing a unique opportunity to study the chemical composition of the nebula at a level normally unachievable in another emission line nebulae [9,10].

Saloman [11] published a revised compilation of energy levels and observed spectral lines of all ionization stages of Xe, referring to studies published to date [12–16]. Light sources include direct-current hollow cathode discharge, theta-pinch discharge, and pulsed capillary discharge. Most of the information is from two studies: Tauheed et al. [13] classified 114 Xe IV lines in VUV using a modified triggered spark initiated by a xenon gas blast as spectral source, and Gallardo et al. [14], who analyzed the 5s²5p²6p, 5s²5p²4f, 5s5p⁴, 5s²5p²5d, and 5s²5p²6s configurations, providing the wavelengths for 618 classified lines in their list, using a capillary discharge as light source.

More recently the study by Raineri et al. [15] reported the weighted oscillator strengths and cancellation factor (CF), calculated from fitted values of the energy parameters of all 769 dipole electric lines belonging to the Xe IV spectrum reported in the compilation [11], including 49 new classified lines.

Hartree–Fock relativistic (HFR) calculations and parametric fits were used. In addition, the results presented in their study were compared to those from Bertuccelli et al. [16].

In order to proceed withthe study of the threetimes ionized xenon spectrum, a new spectral analysis of this ion is presented in this paper. New 36 energy levels for $5s^2 5p^2$ (6d+7s) configurations and 163 new transitions in the 1070–6400 Å region are reported. The relativistic Hartree–Fock method based on the code of Cowan [17] was used. The energy matrix was calculated using energy parameters adjusted to fit the experimental energy levels. Core polarization effects were taken into account in our calculations [18]. We also present a multiconfigurational relativistic approach for the Dirac equation (MCDF), by using the general relativistic atomic structure package (GRASP) [19].

2. Experimental Methods

The spectral source used in this study is based on the pulsed discharge tube built at the Centro de Investigaciones Opticas to study highly ionized noble gases [20]. It consists of a Pyrex tube of about 100 cm with inner diameter of 0.5 cm. The electrodes, placed 80 cm apart, are made of tungsten covered with indium to avoid the impurities coming from the electrodes. The gas excitation was produced by discharging a bank of low-inductance capacitors ranging from 20 to 280 nF, charged with voltages up to 20 kV. The VUV region of the spectrum was recorded using a 3m normal incidence spectrograph equipped with 1200 lines/mm concave diffraction grating and with a plate factor of 2.77 Å/mm in the first diffraction order. Internal wavelength standards are from C, N, O, and known lines of xenon. The wavelength range above 2000 Å was recorded using a 3.4 m Ebert plane-grating spectrograph with 600 lines/mm and a plate factor of 5 Å/mm in the first diffraction order. Thorium lines from an electrodeless discharge were superimposed on the spectrograms and served as reference lines. A photoelectric semiautomatic Grant comparator was used to measure the spectrograms. The uncertainty of the wavelength values of lines was estimated to be correct to ±0.01Å above 2000 Å and ±0.02 Å in the VUV region.

3. Results and Discussion

In this study, we used the modified version of Cowan's atomic calculation package [17], described in our paper [18], for the inclusion of the polarization potentials as a modification in the Hartree–Fock equations. In addition, we considered the corrections of the reduced matrix element used in our previous papers [21], which is the same modification used by Quinet et al. [22] to correct transition matrix elements when including CP and core penetration effects. These methods demand knowledge on the polarizability andcore cut-off radius. The value of α_d for Xe IV core, that is, for Xe 8+ is given by Koch [23] in 0.81130 $a_0{}^3$ and the rc value in 1.16 a_0, defines the boundaries of the atomic core.

We adjusted the values of energy parameters to the experimental energy levels of the Xe IV through a least-squares calculation. With the adjusted values, we calculated the composition of the $5s^2 5p^2$ (6d+7s) energy levels presented in Table 1, where we included lifetimes calculated using HFR and HFR+CP with adjusted energy parameters (here named HFRa and HFR+CPa, respectively) and using multiconfigurational Dirac Fock (MCDF). The MCDF approach was carried out with the extended average level assuming a uniform charge distribution in the nucleus, with a xenon atomic weight of 131.3. The values presented in this work for lifetimes in the MCDF calculation are in Babushkin gauge since this one, in the non-relativistic limits (length), has been found to be the most stable value in many situations, in the sense that it converges smoothly as more correlation is included [24].

Table 1. Energy levels, composition, and lifetimes of Xe IV.

Designation	Energy (cm⁻¹)		Composition	Lifetime(ns)		
	Exp.	Fitted		HFRa	HFR +CPa	MCDF Babushkin
$5s^25p^2(^3P)7s$ $^4P_{1/2}$	239,145	239,126	68.8%^{4}P + 22.5% $5s^25p^2(^3P)7s$ ^{2}P + 8.1% $5s^25p^2(^1S)7s$ ^{2}S	0.304	0.297	0.364
$^4P_{3/2}$	246,689	246,769	84.5%^{4}P + 9.2% $5s^25p^2(^3P)7s$ ^{2}P + 3.3% $5s^25p^2(^3P)6d$ ^{4}D	0.324	0.322	0.414
$^2P_{1/2}$	247,583	247,559	66.7%^{2}P + 23.2% $5s^25p^2(^3P)7s$ ^{4}P + 3.8% $5s^25p^2(^3P)6d$ ^{2}P	0.247	0.232	0.298
$^4P_{5/2}$	251,851	251,784	74.6%^{4}P + 22.9% $5s^25p^2(^1D)7s$ ^{2}D	0.355	0.357	0.553
$^2P_{3/2}$	252,943	252,992	62.5%^{2}P + 25.1% $5s^25p^2(^1D)7s$ ^{2}D + 6.7% $5s^25p^2(^3P)7s$ ^{4}P	0.190	0.190	0.287
$5s^25p^2(^1D)7s$ $^2D_{5/2}$	266,331	266,382	60.5%^{2}D + 17.1% $5s^25p^2(^3P)7s$ ^{4}P + 9.6% $5s^25p^2(^1D)6d$ ^{2}D	0.239	0.257	0.287
$^2D_{3/2}$	266,623	266,574	68.9%^{2}D + 22.7% $5s^25p^2(^3P)7s$ ^{2}P + 3.6% $5s^25p^2(^3P)7s$ ^{4}P	0.221	0.228	0.176
$5s^25p^2(^1S)7s$ $^2S_{1/2}$	283,512	283,519	91.3%^{2}S + 5.2% $5s^25p^2(^3P)7s$ ^{4}P + 2.9% $5s^25p^2(^3P)7s$ ^{2}P	0.318	0.302	0.454
$5s^25p^2(^3P)6d$ $^4F_{3/2}$	234,291	234,304	58%^{4}F + 15% $5s^25p^2(^3P)6d$ ^{4}D + 10% $5s^25p^2(^3P)6d$ ^{2}P	0.519	0.563	0.507
$^4F_{5/2}$	235,660	235,710	35.1%^{4}F + 15.3% $5s^25p^2(^3P)6d$ ^{4}P + 28.5% $5s^25p^2(^3P)6d$ ^{4}D	0.369	0.427	0.337
$^2P_{3/2}$	241,896	241,998	31.2%^{2}P + 33.3% $5s^25p^2(^3P)6d$ ^{4}F + 22.7% $5s^25p^2(^3P)6d$ ^{4}D	0.380	0.416	0.230
$^4F_{7/2}$	242,080	242,086	63.1%^{4}F + 30.5% $5s^25p^2(^3P)6d$ ^{4}D + 5.4% $5s^25p^2(^3P)6d$ ^{2}F	0.744	0.796	0.679
$^4D_{1/2}$	242,541	242,397	80.7%^{4}D + 12.6% $5s^25p^2(^3P)6d$ ^{2}P + 5.5% $5s^25p^2(^3P)6d$ ^{4}P	0.527	0.570	0.481
$^4P_{5/2}$	242,534	242,571	28.3%^{4}P + 54.8% $5s^25p^2(^3P)6d$ ^{4}F + 5.1% $5s^25p^2(^3P)6d$ ^{4}D	0.338	0.389	0.302
$^4D_{3/2}$	244,577	244,535	30.2%^{4}D + 35.1% $5s^25p^2(^3P)6d$ ^{2}P + 14.7% $5s^25p^2(^3P)6d$ ^{4}P	0.299	0.330	0.314
$^2F_{5/2}$	244,722	244,609	65.4%^{2}F + 19.4% $5s^25p^2(^3P)6d$ ^{4}P + 9.9% $5s^25p^2(^1D)6d$ ^{2}F	0.277	0.326	0.223
$^4D_{7/2}$	246,494	246,470	36.2%^{4}D + 26.4% $5s^25p^2(^3P)6d$ ^{4}F + 18.9% $5s^25p^2(^1D)6d$ ^{2}F	0.760	0.804	0.793
$^4F_{9/2}$	246,662	246,625	80.2%^{4}F + 19.6% $5s^25p^2(^1D)6d$ ^{2}G	0.845	0.891	0.818
$^4D_{5/2}$	248,027	248,123	49.6%^{4}D + 24.6% $5s^25p^2(^3P)6d$ ^{4}P + 16% $5s^25p^2(^1D)6d$ ^{2}D	0.281	0.325	0.234
$^4P_{3/2}$	248,565	248,623	57.2%^{4}P + 19.7% $5s^25p^2(^3P)6d$ ^{4}D + 12.7% $5s^25p^2(^1D)6d$ ^{2}P	0.259	0.313	0.205
$^4P_{1/2}$	249,115	249,043	75.8%^{4}P + 11.7% $5s^25p^2(^3P)6d$ ^{2}S + 4.9% $5s^25p^2(^1D)6d$ ^{2}P	0.197	0.235	0.159
$^2P_{1/2}$	250,691	250,595	69.6%^{2}P + 10.3% $5s^25p^2(^3P)6d$ ^{4}D + 6.6% $5s^25p^2(^3P)7s$ ^{2}P	0.213	0.244	0.165
$^2F_{7/2}$	251,073	251,083	55.5%^{2}F + 22.9% $5s^25p^2(^1D)6d$ ^{2}G + 17% $5s^25p^2(^3P)6d$ ^{4}D	0.221	0.267	0.176
$^2D_{5/2}$	251,211	251,204	61.9%^{2}D + 25% $5s^25p^2(^1D)6d$ ^{2}F + 7.9% $5s^25p^2(^1D)6d$ ^{2}D	0.203	0.251	0.144
$^2D_{3/2}$	251,890	251,977	59.7%^{2}D + 14.1% $5s^25p^2(^1D)6d$ ^{2}D + 8.3% $5s^25p^2(^3P)6d$ ^{4}D	0.249	0.295	0.148
$5s^25p^2(^1D)6d$ $^2F_{7/2}$	260,362	260,428	58.7%^{2}F + 20.5% $5s^25p^2(^1D)6d$ ^{2}G + 14.1% $5s^25p^2(^3P)6d$ ^{4}D	0.546	0.614	0.527
$^2G_{9/2}$	261,548	261,656	80.3%^{2}G + 19.6% $5s^25p^2(^3P)6d$ ^{4}F	0.839	0.906	0.862
$^2D_{5/2}$	262,379	262,321	44.1%^{2}D + 30.4% $5s^25p^2(^1D)6d$ ^{2}F + 10% $5s^25p^2(^3P)6d$ ^{4}D	0.196	0.230	0.155
$^2D_{3/2}$	262,438	262,480	69.4%^{2}D + 7.5% $5s^25p^2(^1D)6d$ ^{2}P + 6.9% $5s^25p^2(^3P)6d$ ^{4}D	0.206	0.234	0.171
$^2P_{1/2}$	262,937	262,904	85.4%^{2}P + 4.6% $5s^25p^2(^3P)6d$ ^{2}P + 4.3% $5s^25p^2(^3P)6d$ ^{4}D	0.323	0.395	0.286
$^2G_{7/2}$	262,860	262,969	48.4%^{2}G + 22.2% $5s^25p^2(^3P)6d$ ^{2}F + 19.3% $5s^25p^2(^1D)6d$ ^{2}F	0.234	0.293	0.175

Table 1. *Cont.*

Designation		Energy (cm^{-1})		Composition	Lifetime(ns)		
		Exp.	Fitted		HFRa	HFR +CPa	MCDF Babushkin
5s^{2}5p^2(^{1}S)6d	^{2}F$_{5/2}$	265,205	265,171	24.4%^{2}F + 22.1% 5s^{2}5p^2(^{3}P)6d ^{2}D + 20.3% 5s^{2}5p^2(^{1}D)6d ^{2}D	0.216	0.253	0.157
	^{2}P$_{3/2}$	265,501	265,400	64.9%^{2}P + 9.7% 5s^{2}5p^2(^{3}P)6d ^{2}P + 7.4% 5s^{2}5p^2(^{3}P)6d ^{2}D	0.278	0.334	0.346
	^{2}S$_{1/2}$	265,930	265,908	81.3%^{2}S + 11.4% 5s^{2}5p^2(^{3}P)6d ^{4}P + 4.6% 5s^{2}5p^2(^{3}P)6d ^{2}P	0.198	0.226	0.171
	^{2}D$_{5/2}$	280,142	279,777	91.2%^{2}D + 2.8% 5s^{2}5p^2(^{3}P)6d ^{2}F + 2.3% 5s^{2}5p^2(^{3}P)6d ^{4}D	0.361	0.421	0.316
	^{2}D$_{3/2}$	279,799	280,132	89.4%^{2}D + 4.3% 5s^{2}5p^2(^{3}P)6d ^{2}D + 2.8% 5s^{2}5p^2(^{3}P)6d ^{4}F	0.244	0.293	0.211

In the analysis of spectroscopic data, we take into account isoelectronic trends, Ritz combinations, least-squares adjustment, and relative line intensities in order to identify 36 energy levels belonging to $5s^25p^2(6d+7s)$ configurations for the first time.

As for the isoelectronic sequence calculations used to produce the plots for observed minus calculated ("obs.-calc.") trends along the six first elements of the Sb sequence, we used the configurations $5s^25p^3$, $5s^25p^24f$, $5s^25p^26p$, $5s5p^36s$, $5s5p^37s$, $5s5p^35d$, $5s5p^3\,6d$, $5p^5$ for odd parity and $5s5p^4$, $5s^25p^25d$, $5s5p^35f$, $5s^25p^25g$, $5s^25p^26s$, $5s^25p^26d$, $5s^25p^27s$ for even parity. The calculations included core polarization effects (HFR+CP), with the values of α_d and rc taken from Table 2.

It must be noted that we implemented the modifications suggested by Kramida [25,26] to correct an error in Cowan's package in order to perform the calculations presented here.

Table 2. Values for polarizability α_d and cut-off radius r_c, used in antimony isoelectronic sequence calculations (HFR+CP). Here, a_0 is the Bohr radius.

Ion	α_d $(a_0{}^3)$	r_c (a_0)
Sb I	1.61620	1.33000
Te II	1.25140	1.27000
I III	0.99660	1.21000
Xe IV	0.81130	1.16000
Cs V	0.67210	1.11000
Ba VI	0.56500	1.07000

Data for isoelectronic analysis are from NIST [27] for Sb I, Te II, I III and from Sharman, Tauheed, and Rahimullah for Ba VI [28]. Our analysis is synthesized in Figures 1–3. Surely the LS coupling scheme is not the most appropriate to describe the 6d and 7s configurations, which we concluded after glancing over configuration purities; intermediate couplings provide better descriptions for these levels. We observed a strong eigenvector mixing for all elements studied. However, most of the isoelectronic data available for comparisons are described in the LS scheme, and that was the reason why we chose it.

There is no absolute scale for experimental intensity and therefore we only test its proportionality with the theoretical intensity. We do not include corrections due to the variation of plate reflectivity as a function of wavelength—there is no precise model for this. Our criterion for statistical correlation is to obtain a positive value as close as possible to the unit. Therefore, having a good statistical correlation supports our analysis, but it is just one of the analysis criteria.

The formula $I \propto \sigma gA$ from Cowan's book [17], page 403, tells us that line intensity is proportional to wavenumber σ and weighted transition probability. We analyzed the statistical correlation of the logarithm related to this quantity with the experimental line intensities, which is a visual estimate of the plate blackening (hence the logarithm), obtaining 0.20 for the array $6p - 6d$, 0.32 for $6p - 7s$, and 0.34 for $4f - 6d$. These values were acquired by the HFR+CPa calculation, which is close to HFRa and much better thanab initioHFR and HFR+CP calculations. We also performed a MCDF calculation for gA values. Its agreement with the experimental line intensity shows a poor correlation when compared with HFRa and HFR+CPa for $log(\sigma gA)$, that is, 0.06 for the $6p - 6d$ line array, 0.14 for $6p - 7s$, and 0.18 for $4f - 6d$. It is important to note that our MCDF calculations were performed using a non-current version of the GRASP code where more configurations could not be included. By using a newer version of Grasp codes it would be possible to expand the number of configurations to get better results, which could be more competitive to HFRa and HFR + CPa methods

To understand thesignificance of these values, we compared our values of gA with the experimental values that are in the paper by Bertuccelli et al. [16]. Similarly to them, only 25% of our gA values (HFR+CPa) are within the experimental error. However, a statistical correlation of 0.94 indicates that our values are very linearly proportional to their experiment. When considering the same lines of [16], but substituting their experimental gA values by our estimates for line intensity, correlation with HFR+CPa $log(\sigma gA)$ results in 0.33 for the $6s - 6p$ line array, 0.48 for $5d - 6p$, and 0.50 for $5d - 4f$. Therefore, we can conclude that the calculated σ gA values support our line classification with reasonable agreement.

It is important to note that in this spectral analysis all new levels but two are classified on the basis of two or more lines. The level $^4F_{5/2}$ is a classification attempt based on the only possible line in our spectrograms at 1801.53 Å, a transition with 4f:$^4G_{5/2}$, the strongest spontaneous emission from this level. However, this value does not fit the isoelectronic "obs.-calc." curve. We remove this problem by switching the positions of levels 6d:$^4P_{5/2}$ and 6d:$^4F_{5/2}$ for Xe IV in the isoelectronic analysis. An intense mixing for 6d:$^4P_{5/2}$, $^4D_{5/2}$, and $^4F_{5/2}$ makes the components for the eigenvectors exchange their intensity along with the four first elements, and our choice grouped the energy of the respective multiplets.

Due to similar reasons, we also switched $^4D_{5/2}$ and $^4P_{5/2}$ energy levels for Te III and I III in the respective isoelectronic sequences.

The other level that only has one observed transition is (^1S)6d: $^2D_{3/2}$ that we confirm by our isoelectronic analysis and considering the good agreement in the least squares fit calculation.

There is not much data available for isoelectronic analysis. The lack of information on Cesium and the composition mixing makes level designation a challenge. However, the isoelectronic sequences agree reasonably well with our designations.

Table 3 shows 163 Xe IV lines classified for the first time for transitions involving 5s²5p²(6d+7s) energy levels. We also calculated the weighted transition probability rate gA, where g is the statistical weight 2J+1 of the upper level. We presented gA values obtained from the four methods studied: With and without optimized parameters obtained from least-squares calculations, and with and without core polarization effects for wavefunctions and reduced matrix elements calculations.In these methods, we used the same configuration sets as in [15], that is, 5s²5p³, 5²5p² (4f+6p), 5s 5p³5d, 5p⁵ and 5s5p⁴, 5s²5p² (6s+7s+5d+6d) configurations for odd and even parities, respectively.

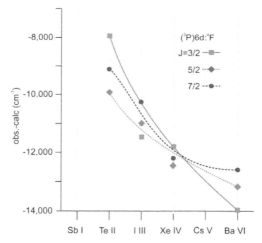

Figure 1. Isoelectronic trend for the multiplet (^3P) ^4F energy levels of the 5s²5p² 6d configuration.

Table 3. Transitions and weighted transition rates for Xe IV.

Int	λ (Å)	Energy (cm⁻¹)		Designation		Weighted Transition Rates—gA (s⁻¹)		
						Adjusted		
		Lower Level	Upper Level	Lower Level	Upper Level	HFRa	HFR+CPa	HFR+CP
1	1078.51	190,793	283,512	$5s^25p^2(^3P)6p\ ^4D_{3/2}$	$5s^25p^2(^1S)7s\ ^2S_{1/2}$	6.623×10^6	3.771×10^6	9.474×10^4
3	1115.44	193,861	283,512	$5s^25p^2(^3P)6p\ ^2S_{1/2}$	$5s^25p^2(^1S)7s\ ^2S_{1/2}$	8.542×10^4	8.067×10^4	2.572×10^5
2	1139.89	195,785	283,512	$5s^25p^2(^3P)4f\ ^4D_{3/2}$	$5s^25p^2(^1S)7s\ ^2S_{1/2}$	2.395×10^4	2.750×10^3	2.061×10^5
2	1151.31	196,655	283,512	$5s^25p^2(^3P)4f\ ^4D_{1/2}$	$5s^25p^2(^1S)7s\ ^2S_{1/2}$	9.396×10^3	6.026×10^3	4.447×10^4
2	1212.39	201,028	283,512	$5s^25p^2(^3P)6p\ ^4S_{3/2}$	$5s^25p^2(^1S)7s\ ^2S_{1/2}$	1.932×10^4	5.698×10^4	7.434×10^6
2	1240.09	182,219	262,860	$5s^25p^2(^3P)4f\ ^4G_{7/2}$	$5s^25p^2(^1D)6d\ ^2G_{7/2}$	1.210×10^6	9.729×10^5	9.768×10^5
2	1259.87	204,140	283,512	$5s^25p^2(^3P)6p\ ^4P_{3/2}$	$5s^25p^2(^1S)7s\ ^2S_{1/2}$	7.200×10^6	6.818×10^6	2.361×10^5
2	1291.10	206,061	283,512	$5s^25p^2(^3P)6p\ ^2P_{3/2}$	$5s^25p^2(^1S)7s\ ^2S_{1/2}$	1.288×10^7	1.543×10^7	1.891×10^7
2	1326.93	189,842	265,205	$5s^25p^2(^3P)4f\ ^4D_{7/2}$	$5s^25p^2(^1D)6d\ ^2F_{5/2}$	3.168×10^7	2.181×10^7	2.609×10^7
2	1340.56	205,205	279,799	$5s^25p^2(^1D)4f\ ^2F_{5/2}$	$5s^25p^2(^1S)6d\ ^2D_{3/2}$	7.092×10^6	1.296×10^6	1.155×10^7
1	1357.65	188,721	262,379	$5s^25p^2(^3P)4f\ ^2D_{5/2}$	$5s^25p^2(^1D)6d\ ^2D_{5/2}$	4.998×10^6	4.123×10^6	3.280×10^6
2	1386.74	188,252	260,362	$5s^25p^2(^3P)4f\ ^4G_{9/2}$	$5s^25p^2(^1D)6d\ ^2F_{7/2}$	1.755×10^6	1.789×10^6	1.046×10^6
1	1394.58	189,842	261,548	$5s^25p^2(^3P)4f\ ^4D_{7/2}$	$5s^25p^2(^1D)6d\ ^2G_{9/2}$	3.174×10^3	7.428×10^3	7.100×10^3
1	1419.25	191,978	262,438	$5s^25p^2(^3P)4f\ ^4D_{5/2}$	$5s^25p^2(^1D)6d\ ^2D_{3/2}$	5.438×10^5	2.925×10^5	2.289×10^5
1	1420.42	191,978	262,379	$5s^25p^2(^3P)4f\ ^4D_{5/2}$	$5s^25p^2(^1D)6d\ ^2D_{5/2}$	4.326×10^6	3.840×10^6	7.217×10^6
1	1430.66	196,725	266,623	$5s^25p^2(^3P)6p\ ^2D_{3/2}$	$5s^25p^2(^1D)7s\ ^2D_{3/2}$	1.565×10^7	9.181×10^6	3.609×10^7
1	1434.39	195,785	265,501	$5s^25p^2(^3P)4f\ ^4D_{3/2}$	$5s^25p^2(^1D)6d\ ^2P_{3/2}$	5.789×10^6	3.652×10^6	2.398×10^5
1	1477.52	198,943	266,623	$5s^25p^2(^3P)6p\ ^4D_{5/2}$	$5s^25p^2(^1D)7s\ ^2D_{3/2}$	1.049×10^6	2.225×10^5	3.155×10^6
1	1496.21	186,109	252,943	$5s^25p^2(^3P)6p\ ^4D_{1/2}$	$5s^25p^2(^3P)7s\ ^2P_{3/2}$	2.673×10^7	3.318×10^7	1.700×10^7
2	1507.08	196,506	262,860	$5s^25p^2(^3P)4f\ ^4F_{5/2}$	$5s^25p^2(^1D)6d\ ^2G_{7/2}$	1.549×10^4	5.596×10^5	9.001×10^6
1	1521.44	200,899	266,623	$5s^25p^2(^3P)6p\ ^4P_{1/2}$	$5s^25p^2(^1D)7s\ ^2D_{3/2}$	2.177×10^5	1.502×10^4	1.926×10^5
9	1551.78	182,219	246,662	$5s^25p^2(^3P)4f\ ^4G_{7/2}$	$5s^25p^2(^3P)6d\,^4F_{9/2}$	1.587×10^7	1.008×10^7	9.882×10^6
4	1552.22	180,152	244,577	$5s^25p^2(^3P)4f\ ^4G_{5/2}$	$5s^25p^2(^3P)6d\,^4D_{3/2}$	8.606×10^5	2.951×10^5	2.936×10^7
7	1573.83	187,533	251,073	$5s^25p^2(^3P)4f\ ^2F_{7/2}$	$5s^25p^2(^3P)6d\,^2F_{7/2}$	1.203×10^7	7.570×10^6	4.904×10^6
7	1615.28	201,028	262,937	$5s^25p^2(^3P)6p\ ^4S_{3/2}$	$5s^25p^2(^1D)6d\ ^2P_{1/2}$	6.055×10^7	4.595×10^7	5.730×10^6
6	1618.41	204,140	265,930	$5s^25p^2(^3P)6p\ ^4P_{3/2}$	$5s^25p^2(^1D)6d\ ^2S_{1/2}$	1.484×10^7	2.801×10^6	8.778×10^5
1	1626.69	186,109	247,583	$5s^25p^2(^3P)6p\ ^4D_{1/2}$	$5s^25p^2(^3P)7s\ ^2P_{1/2}$	2.319×10^7	2.526×10^7	2.055×10^7
3	1645.20	202,076	262,860	$5s^25p^2(^3P)4f\ ^4F_{9/2}$	$5s^25p^2(^1D)6d\ ^2G_{7/2}$	6.376×10^7	4.843×10^7	1.556×10^8
5	1650.75	186,109	246,689	$5s^25p^2(^3P)6p\ ^4D_{1/2}$	$5s^25p^2(^3P)7s\ ^4P_{3/2}$	1.606×10^6	1.586×10^6	2.029×10^6
7	1658.00	182,219	242,534	$5s^25p^2(^3P)4f\ ^4G_{7/2}$	$5s^25p^2(^3P)6d\,^4F_{5/2}$	3.389×10^7	2.586×10^7	3.940×10^8
2	1665.77	191,858	251,890	$5s^25p^2(^3P)4f\ ^4F_{3/2}$	$5s^25p^2(^3P)6d\,^2D_{3/2}$	2.043×10^7	2.421×10^7	7.476×10^5

Table 3. *Cont.*

Int	λ (Å)	Energy (cm^{-1})		Designation		Weighted Transition Rates—gA (s^{-1})		
		Lower Level	Upper Level	Lower Level	Upper Level	HFRa	Adjusted HFR+CPa	HFR+CP
4	1666.86	191,858	251,851	$5s^25p^2(^3P)4f\,^4F_{3/2}$	$5s^25p^2(^3P)7s\,^4P_{5/2}$	1.451×10^6	7.145×10^4	1.065×10^5
5	1670.08	200,486	260,362	$5s^25p^2(^3P)6p\,^2D_{5/2}$	$5s^25p^2(^1D)6d\,^2F_{7/2}$	1.713×10^7	1.995×10^7	7.747×10^7
6	1670.33	206,061	265,930	$5s^25p^2(^3P)6p\,^2P_{3/2}$	$5s^25p^2(^1D)6d\,^2S_{1/2}$	1.361×10^8	1.050×10^8	1.881×10^8
3	1670.60	182,219	242,080	$5s^25p^2(^3P)4f\,^4G_{7/2}$	$5s^25p^2(^3P)6d^4F_{7/2}$	1.019×10^7	7.394×10^6	1.384×10^7
3	1678.74	207,057	266,623	$5s^25p^2(^3P)6p\,^4P_{5/2}$	$5s^25p^2(^1D)7s\,^2D_{3/2}$	1.362×10^8	8.177×10^7	6.031×10^6
5	1681.48	202,076	261,548	$5s^25p^2(^3P)4f\,^4F_{9/2}$	$5s^25p^2(^1D)6d\,^2G_{9/2}$	5.905×10^7	4.145×10^7	2.013×10^7
7	1686.18	188,721	248,027	$5s^25p^2(^3P)4f\,^2D_{5/2}$	$5s^25p^2(^3P)6d^4D_{5/2}$	3.718×10^6	2.399×10^6	3.658×10^6
4	1691.19	187,533	246,662	$5s^25p^2(^3P)4f\,^2G_{7/2}$	$5s^25p^2(^3P)6d^4F_{9/2}$	1.109×10^7	8.797×10^6	2.209×10^7
4	1692.52	193,861	252,943	$5s^25p^2(^3P)6p\,^2S_{1/2}$	$5s^25p^2(^3P)7s\,^2P_{3/2}$	6.667×10^7	7.498×10^7	2.117×10^7
6	1694.50	224,498	283,512	$5s^25p^2(^1D)6p\,^2P_{3/2}$	$5s^25p^2(^1S)7s\,^2S_{1/2}$	3.004×10^6	1.175×10^7	3.405×10^5
7	1709.63	206,713	265,205	$5s^25p^2(^1D)4f\,^2G_{7/2}$	$5s^25p^2(^1D)6d\,^2F_{5/2}$	1.281×10^7	1.748×10^6	2.737×10^7
6	1710.31	186,109	244,577	$5s^25p^2(^3P)6p\,^4D_{1/2}$	$5s^25p^2(^3P)6d^4D_{3/2}$	9.124×10^7	1.199×10^8	1.594×10^8
2	1712.04	188,252	246,662	$5s^25p^2(^3P)4f\,^4G_{9/2}$	$5s^25p^2(^3P)6d^4F_{9/2}$	3.385×10^7	2.261×10^7	2.546×10^7
4	1730.94	188,721	246,494	$5s^25p^2(^3P)4f\,^2D_{5/2}$	$5s^25p^2(^3P)6d^4P_{3/2}$	1.525×10^6	8.100×10^4	1.367×10^6
4	1730.94	190,793	248,565	$5s^25p^2(^3P)6p\,^4D_{3/2}$	$5s^25p^2(^3P)6d^4P_{3/2}$	2.244×10^6	7.523×10^6	2.589×10^6
6	1734.81	205,217	262,860	$5s^25p^2(^1D)4f\,^2F_{7/2}$	$5s^25p^2(^1D)6d\,^2G_{7/2}$	3.324×10^7	2.371×10^7	6.740×10^7
5	1747.19	190,793	248,027	$5s^25p^2(^3P)6p\,^4D_{3/2}$	$5s^25p^2(^3P)6d^4D_{5/2}$	5.268×10^6	7.486×10^6	2.265×10^4
4	1749.08	205,205	262,379	$5s^25p^2(^1D)4f\,^2F_{5/2}$	$5s^25p^2(^1D)6d\,^2D_{5/2}$	8.193×10^7	4.709×10^7	4.908×10^8
4	1749.39	205,217	262,379	$5s^25p^2(^1D)4f\,^2F_{7/2}$	$5s^25p^2(^1D)6d\,^2D_{5/2}$	1.931×10^8	9.089×10^7	1.992×10^8
4	1759.59	193,861	250,691	$5s^25p^2(^3P)6p\,^2S_{1/2}$	$5s^25p^2(^3P)6d^2P_{1/2}$	1.497×10^7	1.798×10^8	1.994×10^8
4	1765.15	189,842	246,494	$5s^25p^2(^3P)4f\,^4P_{7/2}$	$5s^25p^2(^3P)6d^4D_{7/2}$	1.249×10^7	9.511×10^6	2.037×10^7
6	1765.42	206,216	262,860	$5s^25p^2(^1D)4f\,^2H_{9/2}$	$5s^25p^2(^1D)6d\,^2G_{7/2}$	1.762×10^7	1.752×10^7	5.048×10^5
3	1778.77	196,725	252,943	$5s^25p^2(^3P)6p\,^2D_{3/2}$	$5s^25p^2(^3P)7s\,^2P_{3/2}$	9.427×10^7	8.509×10^7	4.613×10^8
4	1780.72	209,344	265,501	$5s^25p^2(^3P)6p\,^2P_{1/2}$	$5s^25p^2(^1D)6d\,^2P_{3/2}$	1.806×10^8	2.079×10^8	4.567×10^8
4	1781.04	206,713	262,860	$5s^25p^2(^1D)4f\,^2G_{7/2}$	$5s^25p^2(^1D)6d\,^2G_{7/2}$	1.404×10^7	1.256×10^7	5.619×10^6
4	1782.39	195,785	251,890	$5s^25p^2(^3P)4f\,^4D_{3/2}$	$5s^25p^2(^3P)6d^2D_{3/2}$	1.465×10^8	1.280×10^8	3.935×10^7
3	1784.18	191,978	248,027	$5s^25p^2(^3P)4f\,^4D_{5/2}$	$5s^25p^2(^3P)6d^4D_{5/2}$	1.175×10^7	6.655×10^6	1.529×10^7
4	1789.02	190,793	246,689	$5s^25p^2(^3P)6p\,^4D_{3/2}$	$5s^25p^2(^3P)7s\,^4P_{3/2}$	1.186×10^8	9.404×10^7	1.597×10^8
4	1797.14	224,498	280,142	$5s^25p^2(^1D)6p\,^2P_{3/2}$	$5s^25p^2(^1S)6d^2D_{5/2}$	5.493×10^7	1.126×10^8	1.487×10^7
4	1800.99	196,325	251,851	$5s^25p^2(^3P)4f\,^4F_{7/2}$	$5s^25p^2(^3P)7s\,^4P_{5/2}$	2.801×10^6	7.025×10^5	3.792×10^5
7	1801.53	180,152	235,660	$5s^25p^2(^3P)4f\,^4G_{5/2}$	$5s^25p^2(^3P)6d^4P_{5/2}$	2.977×10^7	2.086×10^7	8.452×10^6

Table 3. *Cont.*

| Int | λ (Å) | Energy (cm^{-1}) | | Designation | | Weighted Transition Rates—gA (s^{-1}) | | |
		Lower Level	Upper Level	Lower Level	Upper Level	HFRa	Adjusted HFR+CPa	HFR+CP
6	1807.29	206,216	261,548	5s^{2}5p^2(^{1}D)4f ^{2}H$_{9/2}$	5s^{2}5p^2(^{1}D)6d ^{2}G$_{9/2}$	3.229×10^7	2.550×10^7	2.172×10^7
6	1813.02	205,205	260,362	5s^{2}5p^2(^{1}D)4f ^{2}F$_{5/2}$	5s^{2}5p^2(^{1}D)6d ^{2}F$_{7/2}$	2.235×10^7	3.081×10^7	2.006×10^7
3	1813.39	205,217	260,362	5s^{2}5p^2(^{1}D)4f ^{2}F$_{7/2}$	5s^{2}5p^2(^{1}D)6d ^{2}F$_{7/2}$	1.362×10^8	1.150×10^8	1.140×10^8
4	1813.95	196,725	251,851	5s^{2}5p^2(^{3}P)6p ^{2}D$_{3/2}$	5s^{2}5p^2(^{3}P)7s ^{4}P$_{5/2}$	1.231×10^8	9.434×10^7	3.600×10^4
1	1821.29	195,785	250,691	5s^{2}5p^2(^{3}P)4f ^{4}D$_{3/2}$	5s^{2}5p^2(^{3}P)6d ^{2}P$_{1/2}$	2.749×10^7	2.287×10^7	5.120×10^6
8	1822.13	189,842	244,722	5s^{2}5p^2(^{3}P)4f ^{4}D$_{7/2}$	5s^{2}5p^2(^{3}P)6d ^{2}F$_{5/2}$	4.321×10^8	3.081×10^8	3.510×10^8
3	1823.68	206,713	261,548	5s^{2}5p^2(^{1}D)4f ^{2}G$_{7/2}$	5s^{2}5p^2(^{1}D)6d ^{2}G$_{9/2}$	1.275×10^8	1.092×10^8	3.008×10^8
5	1833.29	187,533	242,080	5s^{2}5p^2(^{3}P)4f ^{2}G$_{7/2}$	5s^{2}5p^2(^{3}P)6d ^{4}F$_{7/2}$	5.893×10^7	4.396×10^7	5.539×10^7
3	1853.01	196,725	250,691	5s^{2}5p^2(^{3}P)6p ^{4}D$_{3/2}$	5s^{2}5p^2(^{3}P)6d ^{2}P$_{1/2}$	2.641×10^6	8.155×10^5	6.555×10^7
5	1854.27	190,793	244,722	5s^{2}5p^2(^{3}P)4f ^{2}F$_{5/2}$	5s^{2}5p^2(^{3}P)6d ^{2}F$_{5/2}$	5.464×10^5	1.286×10^6	3.530×10^6
3	1858.13	208,621	262,438	5s^{2}5p^2(^{1}D)4f ^{2}G$_{7/2}$	5s^{2}5p^2(^{1}D)6d ^{2}D$_{3/2}$	1.071×10^5	9.091×10^7	2.376×10^7
3	1863.98	206,713	260,362	5s^{2}5p^2(^{1}D)4f ^{2}G$_{7/2}$	5s^{2}5p^2(^{1}D)6d ^{2}F$_{7/2}$	1.796×10^4	1.370×10^6	2.856×10^7
2	1874.10	188,721	242,080	5s^{2}5p^2(^{3}P)4f ^{4}D$_{5/2}$	5s^{2}5p^2(^{3}P)6d ^{4}F$_{7/2}$	7.040×10^6	7.698×10^6	1.683×10^7
5	1883.46	209,344	262,438	5s^{2}5p^2(^{3}P)6p ^{2}P$_{1/2}$	5s^{2}5p^2(^{1}D)6d ^{2}D$_{3/2}$	4.162×10^7	3.800×10^7	7.917×10^7
6	1890.04	198,943	251,851	5s^{2}5p^2(^{3}P)6p ^{4}D$_{5/2}$	5s^{2}5p^2(^{3}P)7s ^{4}P$_{5/2}$	4.565×10^8	3.836×10^8	2.559×10^6
3	1894.62	195,785	248,565	5s^{2}5p^2(^{3}P)4f ^{4}D$_{3/2}$	5s^{2}5p^2(^{3}P)6d ^{4}P$_{3/2}$	9.841×10^6	6.703×10^6	2.840×10^8
7	1897.88	189,842	242,534	5s^{2}5p^2(^{3}P)4f ^{4}D$_{7/2}$	5s^{2}5p^2(^{3}P)6d ^{4}F$_{5/2}$	4.812×10^7	2.681×10^7	1.772×10^6
5	1901.17	191,978	244,577	5s^{2}5p^2(^{3}P)4f ^{4}D$_{5/2}$	5s^{2}5p^2(^{3}P)6d ^{4}D$_{3/2}$	2.501×10^8	1.785×10^8	1.428×10^4
4	1905.07	199,397	251,890	5s^{2}5p^2(^{3}P)4f ^{2}D$_{3/2}$	5s^{2}5p^2(^{3}P)6d ^{2}D$_{3/2}$	5.527×10^7	5.888×10^7	2.821×10^7
7	1906.20	196,655	249,115	5s^{2}5p^2(^{3}P)4f ^{4}P$_{1/2}$	5s^{2}5p^2(^{3}P)6d ^{4}P$_{1/2}$	9.874×10^8	7.103×10^7	6.996×10^7
6	1913.18	198,943	251,211	5s^{2}5p^2(^{3}P)6p ^{4}D$_{5/2}$	5s^{2}5p^2(^{3}P)6d ^{2}D$_{5/2}$	2.215×10^8	1.343×10^8	8.697×10^6
8	1914.28	189,842	242,080	5s^{2}5p^2(^{3}P)4f ^{4}D$_{7/2}$	5s^{2}5p^2(^{3}P)6d ^{4}F$_{7/2}$	3.153×10^7	2.053×10^7	1.020×10^7
4	1918.27	198,943	251,073	5s^{2}5p^2(^{3}P)6p ^{4}D$_{5/2}$	5s^{2}5p^2(^{3}P)6d ^{2}F$_{7/2}$	3.502×10^8	2.137×10^8	1.380×10^7
5	1929.04	196,725	248,565	5s^{2}5p^2(^{3}P)4f ^{4}D$_{3/2}$	5s^{2}5p^2(^{3}P)6d ^{4}P$_{3/2}$	3.905×10^8	2.829×10^8	8.032×10^6
2	1930.57	195,785	247,583	5s^{2}5p^2(^{3}P)4f ^{4}D$_{3/2}$	5s^{2}5p^2(^{3}P)7s ^{2}P$_{1/2}$	1.205×10^8	1.336×10^8	5.103×10^8
4	1960.89	215,626	266,623	5s^{2}5p^2(^{1}D)6p ^{2}F$_{5/2}$	5s^{2}5p^2(^{1}D)7s ^{2}D$_{3/2}$	1.521×10^9	1.359×10^9	9.246×10^8
6	1961.19	200,899	251,890	5s^{2}5p^2(^{3}P)6p ^{4}P$_{1/2}$	5s^{2}5p^2(^{3}P)6d ^{2}D$_{3/2}$	1.312×10^7	1.068×10^7	3.950×10^7
2	1966.19	196,725	247,583	5s^{2}5p^2(^{3}P)6p ^{2}D$_{3/2}$	5s^{2}5p^2(^{3}P)7s ^{2}P$_{1/2}$	5.960×10^8	5.444×10^8	4.223×10^8
2	1967.55	201,028	251,851	5s^{2}5p^2(^{3}P)6p ^{4}S$_{3/2}$	5s^{2}5p^2(^{3}P)7s ^{4}P$_{5/2}$	7.882×10^7	1.314×10^8	1.316×10^9
7	1972.35	232,811	283,512	5s^{2}5p^2(^{1}S)6p ^{2}P$_{1/2}$	5s^{2}5p^2(^{1}S)7s ^{2}S$_{1/2}$	8.420×10^8	7.097×10^8	7.885×10^8
2	1973.04	191,858	242,541	5s^{2}5p^2(^{3}P)4f ^{4}F$_{3/2}$	5s^{2}5p^2(^{3}P)6d ^{4}D$_{1/2}$	1.839×10^8	1.453×10^8	1.563×10^8

Table 3. *Cont.*

Int	λ (Å)	Energy (cm^{-1})		Designation		Weighted Transition Rates—gA (s^{-1})		
		Lower Level	Upper Level	Lower Level	Upper Level	HFRa	Adjusted HFR+CPa	HFR+CP
2	1976.77	200,486	251,073	$5s^25p^2(^3P)6p\ ^2D_{5/2}$	$5s5p^2(^3P)6d\,^2F_{7/2}$	1.277×10^9	1.374×10^9	2.637×10^8
5	1980.87	216,141	266,623	$5s^25p^2(^1D)6p\ ^2D_{3/2}$	$5s^25p^2(^1D)7s\,^2D_{3/2}$	2.008×10^8	2.357×10^8	8.700×10^7
5	2007.72	200,899	250,691	$5s^25p^2(^3P)6p\ ^4P_{1/2}$	$5s^25p^2(^3P)6d\,^2P_{1/2}$	1.785×10^8	1.802×10^8	2.370×10^8
2	2010.79	199,397	249,115	$5s^25p^2(^3P)4f\ ^2D_{3/2}$	$5s^25p^2(^3P)6d\,^4P_{1/2}$	1.249×10^7	4.943×10^6	4.035×10^7
8	2014.59	198,943	248,565	$5s^25p^2(^3P)6p\ ^4D_{5/2}$	$5s5p^2(^3P)6d\,^4P_{3/2}$	3.953×10^7	1.658×10^7	7.665×10^7
1	2016.33	215,626	265,205	$5s^25p^2(^1D)6p\ ^2F_{5/2}$	$5s^25p^2(^1D)6d\,^2F_{5/2}$	8.577×10^8	4.038×10^8	6.030×10^8
3	2022.77	216,911	266,331	$5s^25p^2(^1D)6p\ ^2D_{5/2}$	$5s^25p^2(^1D)7s\,^2D_{5/2}$	5.091×10^8	3.580×10^8	5.529×10^8
4	2025.24	216,141	265,501	$5s^25p^2(^1D)6p\ ^2D_{3/2}$	$5s^25p^2(^1D)6d\,^2P_{3/2}$	9.701×10^7	4.608×10^7	3.323×10^7
2	2033.21	199,397	248,565	$5s^25p^2(^3P)4f\ ^2D_{3/2}$	$5s^25p^2(^3P)6d\,^4P_{3/2}$	2.115×10^5	1.036×10^6	2.923×10^7
1	2036.36	217,240	266,331	$5s^25p^2(^1D)6p\ ^2F_{7/2}$	$5s^25p^2(^1D)7s\,^2D_{5/2}$	2.256×10^9	1.670×10^9	1.307×10^9
1	2036.71	198,943	248,027	$5s^25p^2(^3P)6p\ ^4D_{5/2}$	$5s^25p^2(^3P)6d\,^4D_{5/2}$	6.800×10^8	5.262×10^8	4.763×10^7
2	2048.41	204,140	252,943	$5s^25p^2(^3P)6p\ ^4P_{3/2}$	$5s^25p^2(^3P)7s\,^2P_{3/2}$	1.331×10^6	2.474×10^6	4.866×10^7
2	2071.42	202,951	251,211	$5s^25p^2(^3P)6p\ ^4D_{7/2}$	$5s^25p^2(^3P)6d\,^2D_{5/2}$	3.115×10^7	2.743×10^7	1.306×10^8
5	2071.80	196,325	244,577	$5s^25p^2(^3P)4f\ ^4F_{7/2}$	$5s^25p^2(^3P)6d\,^4D_{3/2}$	1.306×10^8	1.306×10^8	1.306×10^8
7	2073.30	196,506	244,722	$5s^25p^2(^3P)4f\ ^4F_{5/2}$	$5s^25p^2(^3P)6d\,^2F_{5/2}$	2.372×10^7	1.778×10^7	7.477×10^7
7	2073.30	200,899	249,115	$5s^25p^2(^3P)6p\ ^4P_{1/2}$	$5s^25p^2(^3P)6d\,^4P_{1/2}$	8.505×10^7	7.698×10^7	7.506×10^7
3	2074.74	186,109	234,291	$5s^25p^2(^3P)6p\ ^4D_{1/2}$	$5s^25p^2(^3P)6d\,^4F_{3/2}$	4.587×10^9	4.423×10^9	4.385×10^9
3	2077.37	202,951	251,073	$5s^25p^2(^3P)6p\ ^4D_{7/2}$	$5s^25p^2(^3P)6d\,^2F_{7/2}$	3.747×10^8	3.835×10^8	5.358×10^8
3	2078.87	201,028	249,115	$5s^25p^2(^3P)6p\ ^4S_{3/2}$	$5s^25p^2(^3P)6d\,^4P_{1/2}$	8.110×10^7	1.455×10^8	2.137×10^9
9	2079.23	200,486	248,565	$5s^25p^2(^3P)6p\ ^2D_{5/2}$	$5s^25p^2(^3P)6d\,^4P_{3/2}$	1.066×10^9	1.029×10^9	1.502×10^7
3	2081.10	193,861	241,896	$5s^25p^2(^3P)6p\ ^2S_{1/2}$	$5s^25p^2(^3P)6d\,^2P_{3/2}$	2.500×10^9	2.376×10^9	9.586×10^8
2	2093.75	198,943	246,689	$5s^25p^2(^3P)6p\ ^4D_{5/2}$	$5s^25p^2(^3P)7s\,^4P_{3/2}$	1.854×10^9	1.981×10^9	2.738×10^8
2	2094.11	205,205	252,943	$5s^25p^2(^1D)4f\ ^2F_{5/2}$	$5s^25p^2(^3P)7s\,^2P_{3/2}$	3.332×10^7	3.297×10^7	3.947×10^8
6	2102.94	201,028	248,565	$5s^25p^2(^3P)6p\ ^4S_{3/2}$	$5s^25p^2(^3P)6d\,^4P_{3/2}$	9.059×10^7	1.802×10^8	2.927×10^9
2	2136.22	216,141	262,937	$5s^25p^2(^1D)6p\ ^2D_{3/2}$	$5s^25p^2(^1D)6d\,^2P_{1/2}$	8.568×10^8	8.495×10^8	6.086×10^8
2	2147.31	201,028	247,583	$5s^25p^2(^3P)6p\ ^4S_{3/2}$	$5s^25p^2(^3P)7s\,^2P_{1/2}$	5.096×10^8	5.571×10^8	5.606×10^5
4	2149.87	219,002	265,501	$5s^25p^2(^1D)4f\ ^2D_{5/2}$	$5s^25p^2(^1D)6d\,^2P_{3/2}$	1.852×10^8	1.390×10^8	3.914×10^7
2	2183.24	206,061	251,851	$5s^25p^2(^3P)6p\ ^2P_{3/2}$	$5s^25p^2(^3P)7s\,^4P_{5/2}$	2.778×10^7	7.787×10^6	1.668×10^8
2	2183.24	200,899	246,689	$5s^25p^2(^3P)6p\ ^4P_{1/2}$	$5s^25p^2(^3P)7s\,^4P_{3/2}$	6.598×10^8	6.667×10^8	1.009×10^9
3	2207.67	193,861	239,145	$5s^25p^2(^3P)6p\ ^2S_{1/2}$	$5s^25p^2(^3P)7s\,^4P_{1/2}$	4.086×10^6	6.405×10^6	7.185×10^6
1	2214.68	217,240	262,379	$5s^25p^2(^1D)6p\ ^2F_{7/2}$	$5s^25p^2(^1D)6d\,^2D_{5/2}$	1.526×10^8	9.614×10^7	2.354×10^8

Table 3. Cont.

Int	λ (Å)	Energy (cm^{-1})		Designation		Weighted Transition Rates—gA (s^{-1})		
		Lower Level	Upper Level	Lower Level	Upper Level	Adjusted		HFR+CP
						HFRa	HFR+CPa	
1	2239.94	206,061	250,691	$5s^25p^2(^3P)6p\ ^2P_{3/2}$	$5s^25p^2(^3P)6d\,^2P_{1/2}$	2.758×10^8	2.702×10^8	3.217×10^8
1	2242.39	235,561	280,142	$5s^25p^2(^1S)6p\ ^2P_{3/2}$	$5s^25p^2(^1S)6d\,^2D_{5/2}$	6.465×10^9	6.274×10	6.193×10^9
5	2295.89	202,951	246,494	$5s^25p^2(^3P)6p\ ^4D_{7/2}$	$5s^25p^2(^3P)6d\,^4D_{7/2}$	2.371×10^9	2.357×10^9	2.815×10^9
1	2298.23	190,793	234,291	$5s^25p^2(^3P)6p\ ^4D_{3/2}$	$5s^25p^2(^3P)6d\,^4F_{3/2}$	4.862×10^8	4.099×10^8	5.482×10^8
1	2317.55	199,397	242,534	$5s^25p^2(^3P)4f\ ^2D_{3/2}$	$5s^25p^2(^3P)6d\,^4F_{5/2}$	3.469×10^8	4.846×10^8	1.949×10^5
3	2407.63	206,061	247,583	$5s^25p^2(^3P)6p\ ^2P_{3/2}$	$5s^25p^2(^3P)7s\ ^2P_{1/2}$	1.076×10^8	9.367×10^7	7.285×10^7
6	2408.41	207,057	248,565	$5s^25p^2(^3P)6p\ ^4P_{5/2}$	$5s^25p^2(^3P)6d\,^4P_{3/2}$	2.324×10^8	2.041×10^8	1.291×10^9
2	2472.25	204,140	244,577	$5s^25p^2(^3P)6p\ ^4P_{3/2}$	$5s^25p^2(^3P)6d\,^4D_{3/2}$	2.832×10^7	5.368×10^7	1.035×10^8
3	2498.99	202,076	242,080	$5s^25p^2(^3P)4f\ ^4F_{9/2}$	$5s^25p^2(^3P)6d\,^4F_{7/2}$	6.998×10^6	4.645×10^6	9.367×10^5
3	2502.73	208,621	248,565	$5s^25p^2(^3P)4f\ ^2F_{5/2}$	$5s^25p^2(^3P)6d\,^4P_{3/2}$	4.929×10^7	7.502×10^7	5.397×10^5
1	2595.56	206,061	244,577	$5s^25p^2(^3P)6p\ ^2P_{3/2}$	$5s^25p^2(^3P)6d\,^4D_{3/2}$	4.640×10^8	4.162×10^8	4.734×10^7
1	2596.23	195,785	234,291	$5s^25p^2(^3P)4f\ ^4D_{3/2}$	$5s^25p^2(^3P)6d\,^4F_{3/2}$	6.854×10^6	4.678×10^6	1.515×10^6
1	2603.52	204,140	242,541	$5s^25p^2(^3P)6p\ ^4P_{3/2}$	$5s^25p^2(^3P)6d\,^4D_{1/2}$	1.003×10^7	1.586×10^7	4.086×10^7
2	2622.74	201,028	239,145	$5s^25p^2(^3P)6p\ ^4S_{3/2}$	$5s^25p^2(^3P)7s\ ^4P_{1/2}$	1.124×10^7	6.599×10^6	5.223×10^6
1	2789.76	206,061	241,896	$5s^25p^2(^3P)6p\ ^2P_{3/2}$	$5s^25p^2(^3P)6d\,^2P_{3/2}$	4.024×10^7	3.675×10^7	3.847×10^8
4	2855.73	204,140	239,145	$5s^25p^2(^3P)6p\ ^4P_{3/2}$	$5s^25p^2(^3P)7s\ ^4P_{1/2}$	3.366×10^6	1.750×10^6	2.920×10^7
4	3021.77	206,061	239,145	$5s^25p^2(^3P)6p\ ^2P_{3/2}$	$5s^25p^2(^3P)7s\ ^4P_{1/2}$	4.846×10^6	5.063×10^6	5.461×10^6
1	3031.95	216,141	249,115	$5s^25p^2(^1D)6p\ ^2D_{3/2}$	$5s^25p^2(^3P)6d\,^4P_{1/2}$	3.455×10^6	1.022×10^6	7.316×10^4
1	3083.27	216,141	248,565	$5s^25p^2(^1D)6p\ ^2D_{3/2}$	$5s^25p^2(^3P)6d\,^4P_{3/2}$	5.342×10^6	1.452×10^6	2.771×10^5
1	3117.20	219,002	251,073	$5s^25p^2(^1D)4f\ ^2D_{5/2}$	$5s^25p^2(^3P)6d\,^2F_{7/2}$	7.948×10^7	5.664×10^7	4.145×10^6
4	3143.02	220,082	251,890	$5s^25p^2(^1D)6p\ ^2P_{1/2}$	$5s^25p^2(^3P)6d\,^2D_{3/2}$	3.672×10^7	4.216×10^6	7.089×10^7
3	3214.51	220,790	251,890	$5s^25p^2(^1D)4f\ ^2P_{1/2}$	$5s^25p^2(^3P)6d\,^2F_{5/2}$	9.984×10^6	4.387×10^7	2.031×10^6
3	3238.65	215,626	246,494	$5s^25p^2(^1D)6p\ ^2F_{5/2}$	$5s^25p^2(^3P)6d\,^4D_{7/2}$	1.314×10^5	2.120×10^5	1.061×10^6
1	3241.45	213,736	244,577	$5s^25p^2(^1D)4f\ ^2D_{3/2}$	$5s^25p^2(^3P)6d\,^4D_{3/2}$	4.030×10^6	4.550×10^6	1.601×10^5
1	3247.11	217,240	248,027	$5s^25p^2(^1D)6p\ ^2F_{7/2}$	$5s^25p^2(^3P)6d\,^4D_{5/2}$	2.611×10^6	1.551×10^6	2.332×10^5
1	3248.98	235,561	266,331	$5s^25p^2(^1S)6p\ ^2P_{3/2}$	$5s^25p^2(^1D)7s\ ^2D_{5/2}$	2.561×10^6	5.870×10^6	1.354×10^6
2	3515.63	216,141	244,577	$5s^25p^2(^1D)6p\ ^2D_{3/2}$	$5s^25p^2(^3P)6d\,^4D_{3/2}$	1.258×10^4	8.724×10^1	1.223×10^5
2	3550.01	213,736	241,896	$5s^25p^2(^1D)4f\ ^2D_{3/2}$	$5s^25p^2(^3P)6d\,^2P_{3/2}$	7.120×10^5	5.240×10^5	1.064×10^6
2	3594.60	216,911	244,722	$5s^25p^2(^1D)6p\ ^2D_{5/2}$	$5s^25p^2(^3P)6d\,^2F_{5/2}$	8.503×10^5	1.321×10^6	4.517×10^6
2	3636.34	219,002	246,494	$5s^25p^2(^1D)4f\ ^2D_{5/2}$	$5s^25p^2(^3P)6d\,^4D_{7/2}$	6.194×10^6	3.751×10^6	1.249×10^6
1	3637.66	217,240	244,722	$5s^25p^2(^1D)6p\ ^2F_{7/2}$	$5s^25p^2(^3P)6d\,^2F_{5/2}$	4.573×10^5	2.214×10^5	1.336×10^6

Table 3. *Cont.*

| Int | λ (Å) | Energy (cm^{-1}) | | Designation | | Weighted Transition Rates—gA (s^{-1}) | | |
		Lower Level	Upper Level	Lower Level	Upper Level	HFRa	Adjusted HFR+CPa	HFR+CP
3	3654.96	224,498	251,851	$5s^25p^2(^1D)6p\ ^2P_{3/2}$	$5s^25p^2(^3P)7s\ ^4P_{5/2}$	3.333×10^5	9.256×10^4	3.850×10^4
2	3715.25	215,626	242,534	$5s^25p^2(^1D)6p\ ^2F_{5/2}$	$5s^25p^2(^3P)6d\ ^4F_{5/2}$	1.159×10^5	1.184×10^4	4.643×10^5
2	3901.70	216,911	242,534	$5s^25p^2(^1D)6p\ ^2D_{5/2}$	$5s^25p^2(^3P)6d\ ^4F_{5/2}$	2.684×10^5	2.945×10^5	3.263×10^4
4	4061.12	224,498	249,115	$5s^25p^2(^1D)6p\ ^2P_{3/2}$	$5s^25p^2(^3P)6d\ ^4P_{1/2}$	9.721×10^5	4.089×10^5	7.330×10^4
3	4248.40	219,002	242,534	$5s^25p^2(^1D)4f\ ^2D_{5/2}$	$5s^25p^2(^3P)6d\ ^4F_{5/2}$	2.127×10^4	6.785×10^4	2.765×10^2
2	4470.40	219,717	242,080	$5s^25p^2(^3P)4f\ ^2F_{7/2}$	$5s^25p^2(^3P)6d\ ^4F_{7/2}$	8.612×10^5	9.968×10^5	1.711×10^4
1	4366.60	219,002	241,896	$5s^25p^2(^1D)4f\ ^2D_{5/2}$	$5s^25p^2(^3P)6d\ ^2P_{3/2}$	4.677×10^5	3.493×10^5	2.294×10^5
1	4505.10	224,498	246,689	$5s^25p^2(^1D)6p\ ^2P_{3/2}$	$5s^25p^2(^3P)7s\ ^4P_{3/2}$	6.198×10^2	2.209×10^4	4.401×10^4
2	4582.85	220,082	241,896	$5s^25p^2(^1D)6p\ ^2P_{1/2}$	$5s^25p^2(^3P)6d\ ^2P_{3/2}$	5.479×10^5	5.230×10^5	5.416×10^4
1	5240.06	232,811	251,890	$5s^25p^2(^1S)6p\ ^2P_{1/2}$	$5s^25p^2(^3P)6d\ ^2D_{3/2}$	1.262×10^6	1.642×10^6	2.573×10^5
4	6348.69	228,975	244,722	$5s^25p^2(^1S)4f\ ^2F_{7/2}$	$5s^25p^2(^3P)6d\ ^2F_{5/2}$	9.962×10^3	9.580×10^3	3.369×10^3

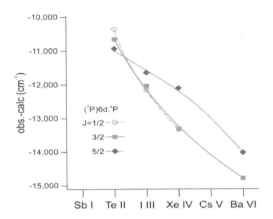

Figure 2. Isoelectronic trend for the multiplet (^{3}P) ^{4}P energy levels of the $5s^25p^2$ 6d configuration.

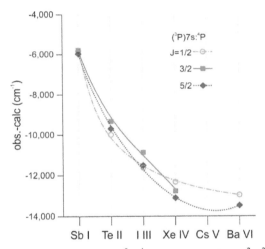

Figure 3. Isoelectronic trend for the multiplet (^{3}P) ^{4}P energy levels of the $5s^25p^2$ 7s configuration.

Table 4 shows the result of least squares adjustment for even parity levels, where 6d and 7s configurations are included. All single configuration parameters, the R^k integrals for $5s5p^4$-$5s^25p^2$6s, $5s5p^4$-$5s^25p^25d$, $5s^25p^26s$-$5s^25p^2$ 5d interactions, and the $R^1(5p,5d;6d,5p)$of the $5s^25p^2$ 5d-$5s^25p^2$6d interaction were left free during the final calculation. The rest of the configuration interaction integrals remained fixed at 85% of their Hartree–Fock values. We found a standard deviation of 138 cm^{-1} for this adjustment.

Table 4. Least-squares parameters for even parity of Xe IV. Standard deviation is 138 cm^{-1}.

Configuration	Parameter	HFR (cm^{-1})		HFRa./HFR [a]
		HFR	HFRa	
$5s5p^4$	$E_{av}(5s5p^4)$	145,275	132,757	−12,519
	$F^2(5p,5p)$	53,464	46,502	87%
	α	0	−402	
	ζ_{5p}	8246	8600	104%
	$G^1(5s,5p)$	70,216	48,430	69%
$5s^25p^26s$	$E_{av}(5s^25p^26s)$	187,245	176,036	−11,209
	$F^2(5p,5p)$	54,783	43,692	80%
	α	0	−55	
	ζ_{5p}	8859	8945	101%
	$G^1(5p,6s)$	5898	4379	74%
$5s^25p^27s$	$E_{av}(5s^25p^27s)$	267,957	257,041	−10,916
	$F^2(5p,5p)$	55,283	47,384	86%
	ζ_{5p}	8999	8556	95%
	$G^1(5p,7s)$	1801	1633	91%
$5s^25p^25d$	$E_{av}(5s^25p^25d)$	170,438	158,790	−11,648
	$F^2(5p,5p)$	54,191	42,089	78%
	α	0	−123	
	ζ_{5p}	8593	8754	102%
	ζ_{5d}	478	695	145%
	$F^2(5p,5d)$	39,705	32,721	82%
	$G^1(5p,5d)$	44,921	32,124	72%
	$G^3(5p,5d)$	28,247	20,111	71%
$5s^25p^26d$	$E_{av}(5s^25p^26d)$	264,034	253,060	−10,975
	$F^2(5p,5p)$	55,267	47,585	86%
	ζ_{5p}	8972	8449	94%
	ζ_{6d}	161	153	95%
	$F^2(5p,6d)$	11,723	10,009	85%
	$G^1(5p,6d)$	7747	6753	87%
	$G^3(5p,6d)$	5444	5575	102%
$5s5p^4$-$5s^25p^26s$	$R^1(5p,5p;5s,6s)$	−1237	−851	69%
$5s5p^4$-$5s^25p^27s$	$R^1(5p,5p;5s,7s)$	−1351	−1148	85%
$5s5p^4$-$5s^25p^25d$	$R^1(5p,5p;5s,5d)$	53,926	37,094	69%
$5s5p^4$-$5s^25p^26d$	$R^1(5p,5p;5s,6d)$	22,435	19,069	85%
$5s^25p^26s$-$5s^25p^27s$	$R^1(5p,6s;7s,5p)$	3120	2652	85%
$5s^25p^26s$-$5s^25p^25d$	$R^2(5p,6s;5p,5d)$	−12,799	−10,336	81%
	$R^1(5p,6s;5d,5p)$	−5075	−4098	81%
$5s^25p^26s$-$5s^25p^26d$	$R^2(5p,6s;5p,6d)$	4779	4062	85%
	$R^1(5p,6s;6d,5p)$	85	73	85%
$5s^25p^27s$-$5s^25p^25d$	$R^2(5p,7s;5p,5d)$	−6519	−5541	85%
	$R^1(5p,7s;5d,5p)$	−3294	−2800	85%
$5s^25p^27s$-$5s^25p^26d$	$R^2(5p,7s;5p,6d)$	−3058	−2599	85%
	$R^1(5p,7s;6d,5p)$	−391	−333	85%
$5s^25p^25d$-$5s^25p^26d$	$R^2(5p,5d;5p,6d)$	12,162	10,338	85%
	$R^1(5p,5d;6d,5p)$	17,415	13,061	75%
	$R^3(5p,5d;6d,5p)$	11,432	9717	85%

[a] Ratio HFRa to HFR for each case, except for average energies, where values are the difference of HFRa minus HFR for each case.

4. Conclusions

In this study we extended the knowledge of the Xe IV spectrum to the $5s^25p^27s$ and $5s^25p^2\,6d$ configuration, from a set of 163 new line classifications. To produce this new information, we used a set of different analysis tools, including calculations from three models (HFR, HFR+CP, and MCDF), least-squares adjustment, line intensity comparisons, and isoelectronic analysis, which makes us very confident in our results.

Author Contributions: All authors contributed equally to this work.

Funding: This research received no external funding.

Acknowledgments: This research was supported by the Consejo Nacional de Investigaciones Científicas y Tecnicas (CONICET), Argentina, and by the Coordenação de Aperfeiçoamento de Pessoal de Nível Superior (CAPES), Brazil, Finance Code 001. The authors thank Espaço da Escrita–Pró-Reitoria de Pesquisa–UNICAMP-for the language services provided. Support of the Comision de Investigaciones Científicas de la Província de Buenos Aires (CIC), where M.R. is a researcher, is also gratefully acknowledged.

Conflicts of Interest: The authors declare no conflicts of interest.

References

1. Gustafsson, B. The future of stellar spectroscopy and its dependence on YOU. *Phys. Scr.* **1991**, *34*, 14–19. [CrossRef]

2. Biémont, E.; Blagoev, K.; Campos, J.; Mayo, R.; Malcheva, G.; Ortíz, M.; Quinet, P. Radiative parameters for some transitions in Cu(II) and Ag(II) spectrum. *Spectrosc. Relat. Phenom.* **2005**, *144*, 27–28. [CrossRef]

3. Cowley, C.R.; Hubrig, S.; Palmeri, P.; Quinet, P.; Biémont, É.; Wahlgren, G.M.; Schütz, O.; González, J.F. HD 65949: Rosetta stone or red herring. *Mon. Not. R. Astron. Soc.* **2010**, *405*, 1271–1284. [CrossRef]

4. Otsuka, M.; Tajitsu, A. Chemical abundances in the extremely carbon-rich and xenon-rich halo planetary nebula H4-1. *Astrophys. J.* **2013**, *778*, 146. [CrossRef]

5. Péquignot, D.; Baluteau, J.-P. The identification of krypton, xenon, and other elements of rows 4, 5 and 6 of the periodic table in the planetary nebula NGC 7027. *Astron. Astrophys.* **1994**, *283*, 593–625.

6. Biémont, E.; Hansen, J.E.; Quinet, P.; Zeippen, C.J. Forbidden transitions of astrophysical interest in the 5pk (k=1–5) *configurations*. *Astron. Astrophys. Suppl. Ser.* **1995**, *111*, 333–346.

7. Werner, K.; Rauch, T.; Ringat, E.; Kruk, J.W. First detection of krypton and xenon in a white dwarf. *Astrophys. J.* **2012**, *753*, L7. [CrossRef]

8. Rauch, T.; Hoyer, D.; Quinet, P.; Gallardo, M.; Raineri, M. The Xe VI ultraviolet spectrum and the xenon abundance in the hot do-type white dwarf RE 0503−289. *Astron. Astrophys.* **2015**, *577*, A88. [CrossRef]

9. Zhang, Y.; Liu, X.-W.; Luo, S.-G.; Péquignot, D.; Barlow, M.J. Integrated spectrum of the planetary nebula NGC 7027. *Astron. Astrophys.* **2005**, *442*, 249–262. [CrossRef]

10. Zhang, Y.; Williams, R.; Pellegrini, E.; Cavagnolo, K.; Baldwin, J.A.; Sharpee, B.; Phillips, M.; Liu, X.-W. Abundances of s-process elements in planetary nebulae: Br, Kr &Xe in Planetary Nebulae in Our Galaxy and Beyond. *Proc. IAU Symp.* **2006**, *234*, 549–550. [CrossRef]

11. Saloman, E.B. Energy levels and observed spectral lines of xenon, Xe I through Xe LIV. *J. Phys. Chem. Ref. Data* **2004**, *33*, 765–921. [CrossRef]

12. Gallardo, M.; ReynaAlmandos, J.G. *XenonLines in the Range from 2000 Å to 7000 Å*; Serie "Monografias Cientificas" No. 1; Centro de Investigaciones Opticas: LaPlata, Argentina, 1981.

13. Tauheed, A.; Joshi, Y.N.; Pinnington, E.H. Revised and extended analysis of the $5s^25p^3$, $5s5p^4$, $5s^25p^25d$ and $5s^25p^26s$ configurations of trebly ionized xenon (Xe IV). *Phys. Scr.* **1993**, *47*, 555–560. [CrossRef]

14. Gallardo, M.; Raineri, M.; Reyna Almandos, J.G.; Di Rocco, H.O.; Bertuccelli, D.; Trigueiros, A.G. 5s25p2(6p + 4f) configurations in triply ionized xenon (Xe IV). *Phys. Scr.* **1995**, *51*, 737–751. [CrossRef]

15. Raineri, M.; Lagorio, C.; Padilla, S.; Gallardo, M.; Reyna Almandos, J. Weighted oscillator strengths for the Xe IV spectrum. *At. Data Nucl. Data Tables* **2008**, *94*, 140–159. [CrossRef]

16. Bertuccelli, G.; Di Rocco, H.O.; Iriarte, D.I.; Pomarico, J.A. Experimental Determination of Transition Probabilities of Xe IV; Comparison with Semiempirical Calculations. *Phys. Scr.* **2000**, *62*, 277–281. [CrossRef]

17. Cowan, R.D. *The Theory of Atomic Structure and Spectra*; University of California Press: Berkeley, CA, USA, 1981.

18. Pagan, C.J.B.; Raineri, M.; Gallardo, M.; Reyna Almandos, J. Spectral Analysis and New Visible and Ultraviolet Lines of ArV. *Astron. Astrophys. Suppl. Ser.* **2019**, *242*, 24. [CrossRef]

19. Dyal, K.G.; Grant, I.; Johnson, C.T.; Parpia, F.A.; Plummer, E.P. GRASP: A general-purpose relativistic atomic structure program. *Comput. Phys. Commun.* **1989**, *55*, 425–456. [CrossRef]

20. Reyna Almandos, J.; Bredice, F.; Raineri, M.; Gallardo, M. Spectral analysis of ionized noble gases and implications for astronomy and laser studies. *Phys. Scr.* **2009**, *T134*, 014018. [CrossRef]

21. Raineri, M.; Gallardo, M.; Pagan, C.J.B.; Trigueiros, A.G.; ReynaAlmandos, J. Lifetimes and transition probabilities in KrV. *J. Quant. Spectrosc. Radiat. Transf.* **2012**, *113*, 1612–1627. [CrossRef]

22. Quinet, P.; Palmeri, P.; Biémont, E.; McCurdy, M.M.; Rieger, G.; Pinnington, E.H.; Wickliffe, M.E.; Lawler, J.E. Experimental and theoretical radiative lifetimes, branching fractions and oscillator strengths in Lu II. *Mon. Not. R. Astron. Soc.* **1999**, *307*, 934–940. [CrossRef]

23. Koch, V.; Andrae, D. Static Electric DipolePolarizabilities for Isoelectronic Sequences. *Int. J. Quantum Chem.* **2011**, *111*, 891–903. [CrossRef]

24. Grant, I. *Relativistic Quantum Theory of Atoms and Molecules: Theory and Computation*; Springer: Oxford, UK, 2007.

25. Kramida, A. Configuration interactions of class 11: Na error in Cowan's atomic structure theory. *Comput. Phys.Commun.* **2017**, *215*, 47–48. [CrossRef]

26. Kramida, A. Corrigendum to "Configuration interactions of class 11: Na error in Cowan's atomic structure theory". *Comput. Phys. Commun.* **2018**, *232*, 266–267. [CrossRef]

27. NIST Standard Reference Database 78. Version 5.6. Available online: https://www.nist.gov/pml/atomic-spectra-database (accessed on 31 October 2018).

28. Sharman, M.K.; Tauheed, A.; Rahimullah, K. Spectral analysis of $5s^2 5p^2$ (6p+6d+7s) configurations of Ba VI. *J. Quant. Spectrosc. Radiat. Transf.* **2014**, *142*, 37–48. [CrossRef]

Article

Atomic Data Needs in Astrophysics: The Galactic Center "Scandium Mystery"

Brian Thorsbro

Department of Astronomy and Theoretical Physics, Lund University, Box 43, 221 00 Lund, Sweden;
thorsbro@astro.lu.se

Received: 14 November 2019; Accepted: 14 January 2020; Published: 22 January 2020

Abstract: Investigating the Galactic center offers unique insights into the buildup and history of our Galaxy and is a stepping stone to understand galaxies in a larger context. It is reasonable to expect that the stars found in the Galactic center might have a different composition compared to stars found in the local neighborhood around the Sun. It is therefore quite exciting when recently there were reports of unusual neutral scandium, yttrium, and vanadium abundances found in the Galactic center stars, compared to local neighborhood stars. To explain the scandium abundances in the Galactic center, we turn to recent laboratory measurements and theoretical calculations done on the atomic oscillator strengths of neutral scandium lines in the near infrared. We combine these with measurements of the hyper fine splitting of neutral scandium. We show how these results can be used to explain the reported unusual scandium abundances and conclude that in this respect, the environment of the Galactic center is not that different from the environment in the local neighborhood around the sun.

Keywords: Galactic center; stellar abundances; scandium; hyper fine splitting

1. Introduction

Understanding the formation and evolution of galaxies is one of the important questions in astrophysics today [1]. The Milky Way, our own Galaxy, is central as it is the galaxy we can observe in the greatest detail, both because we want to explore the world we live in and because we can use it to understand galaxies at large, assuming the mediocrity principle, i.e., that we are not special.

Our Galaxy is currently seen as a barred spiral galaxy with multiple stellar populations, often roughly classified as the halo, the thin and thick disks, the bulge, and the Galactic center [1].

Key components to understanding the formation and evolution of the Galaxy are understanding the star formation rate and the mass distribution of the formed stars [2]. The constituents of the Galaxy that are easiest to observe are its stars, and the Galactic evolution can be decoded from its stars by understanding stellar structure and evolution. Fortunately, stellar structure and evolution theories are well developed, and they allow us to model how stars evolve and eventually perish [3]. In particular, we can model what kind of chemical species are synthesized in stars and consequently distributed to the surrounding environment, from which new stars are born.

Combining galactic and stellar evolution models thus demands the study of chemical evolution models that predict the abundances of chemical species in stars at different times and locations [4]. The chemical composition of the photosphere of a star, which is the region of the star from which light escapes, changes during the life of the star as heavier elements settle. However, as the star turns into a red giant towards the end of its life, convection inside the star remixes the material, undoing the settling of the heavy elements. A few light element species produced by the fusion processes in the center of the star are transported up to the photosphere with this convection as well. Apart from that, however, one can approximately assume to observe the composition of the environment in which the star was born by observing the composition of its current photosphere. These observations are done with stellar spectroscopy, and the abundance of chemical species in the photosphere is determined using abundance analysis, which thus enables comparisons between models and observations.

In our study, we focus on the Galactic center, which is interesting as it is an environment that is unique to the Galaxy, particularly because of the presence of the super massive black hole. This could lead to a different evolution of stars compared to the environment in the vicinity of our Sun. In order to be able to observe stars in the Galactic center, one has to turn to very bright giant stars and observe them in the infrared wavelength regions. More energetic light is absorbed and scattered away by the dust lying between the center and the Sun, and thus, it is not possible to observe the Galactic center in the visible light wavelength range.

One chemical species that is of concern here is neutral scandium. In early 2018, it was reported that unusually high amounts of scandium, together with yttrium and vanadium, might be present in the Galactic center stars [5]. A unique scandium abundance in the Galactic center would suggest that the center is a site for a new channel of nucleosynthesis of neutral scandium and possibly other elements. Such a trend is important to understand, especially when observing the centers of far away galaxies, which would be a natural choice to observe first, as the center is the most luminous part of a galaxy.

In this work, we summarize our findings first reported in Thorsbro et al. [6] and discuss the atomic data needs in astrophysics on the basis of this.

2. Results

2.1. Observations

We observed 18 stars in the Galactic center and eight stars located in the solar neighborhood. The exact information about the observations of these stars, as well as how they were analyzed were detailed by Thorsbro et al. [6], Ryde et al. [7], Rich et al. [8]. All of the stars were of similar stellar classification, denominated M giants, which means that they had an effective temperature[1] between 3000 and 4000 K. The solar neighborhood stars were observed as a control group to compare the Galactic center stars. The stars were observed with the Keck II telescope (10 m class telescope), mounted with the NIRSPECspectrograph [9]. The spectra of the stars were recorded in the near infrared wavelength region around 2 microns ($\sim$5000 cm^{-1}) with a resolving power R $= \frac{\lambda}{\Delta\lambda} \simeq 23{,}000$.

In the spring of 2018, Do et al. [5] reported to have found evidence for unusual high scandium abundances in stars in the Galactic center located within a few parsecs from the super massive black hole of our Galaxy. Their observed stars displayed unusually strong scandium line features in their spectra. They argued that modeling the observed spectra with synthetic spectra from first principles was impossible unless one assumed an unusually high scandium abundance. They further compared their observed stars to a star found in a globular cluster, which was a high stellar density environment, and showed that the comparison star did not show the same unusually high abundances. It is worth noting that the comparison star had an effective temperature that was about 800 to 1000 K warmer compared to the their observed Galactic center stars.

In Figure 1, the spectra from our observations are shown as we published them in Thorsbro et al. [6]. Here, it is worth noticing that the scandium line features were strong for stars located in both the Galactic center and in the solar neighborhood. Further, for both environments, the strength of the scandium line features diminished as the temperature increased. We could thus immediately conclude that the environment of the star did not seem to be connected to the strength of the line features. To improve our understanding, we therefore investigated if there could be assumptions in the modeling from first principles that needed to be revisited.

---

[1] effective temperature is defined by the corresponding black-body radiation curve with an equal amount of radiation to the radiation coming from the star.

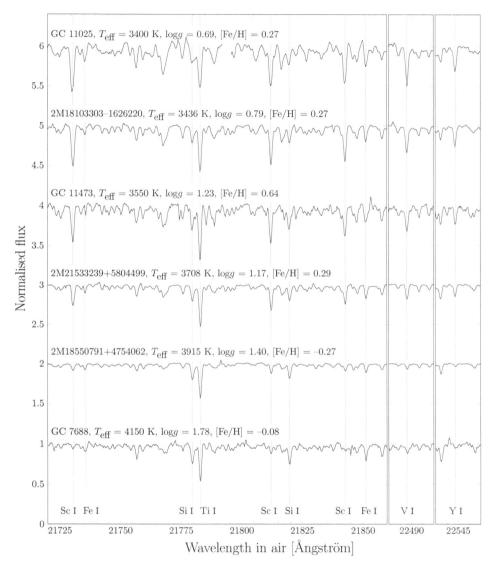

Figure 1. This figure from Thorsbro et al. [6] (their Figure 1) shows that the strong scandium feature seems to be connected to the effective temperature of the star. Six stellar spectra are plotted together with increasing temperature from top to bottom with blue vertical lines identifying lines of interest. The normalized fluxes are translated upward with integer values for presentation. The stars are a mix of three Galactic center stars (GC 7688, GC 11025, and GC 11473) and three solar neighborhood stars (2M18103303–1626220, 2M18550791+4754062, and 2M21533239+5804499). The spectra show strikingly strong scandium, vanadium, and yttrium lines in the cooler stars, even though the stars are located in widely different environments. As temperatures increase to 3900 K and beyond, the neutral lines of scandium, vanadium, and yttrium begin to vanish, presumably due to ionization.

2.2. Analyzing the Hyperfine Structure

The line transition in neutral scandium that is shown in Figure 1 is the $3d^24s$—$3d4s4p$ transition. Neutral scandium has a nuclear spin of $I = 7/2$, and since this transition involves an unpaired 4s-election, the hyperfine splitting is expected to be strong.

In general, there is a lack of experimental atomic data for near-IR transitions [10]. In response to this need, a program was initiated to provide accurate and vetted near-IR atomic data for stellar spectroscopy. This program was lead by Henrik Hartman and Per Jönsson at Malmö University in Sweden using the Edlén laboratory, a joint effort of Malmö and Lund universities.

Recent works measured the oscillator strengths and hyperfine structure of neutral scandium [11,12]. Both of these works utilized the Fourier transform spectrometer in the Edlén laboratory examining neutral scandium energized in a hollow cathode lamp.

In the work of Pehlivan et al. [11] intensity calibrated spectra of Sc I are used to experimentally determine branching fractions, which are then combined with radiative lifetimes from the literature [13–15] to derive accurate oscillator strengths for Sc I.

In the work of van Deelen [12], synthetic model spectra of Sc I hyperfine structure (HFS) multiplets were constructed and fitted to experimentally measured spectra. The results were compared to similar results from the literature [16–23], and investigations were initiated to examine the significant differences more closely. The work of van Deelen [12] can be considered the most recent and accurate compilation of Sc I HFS data for the near infrared wavelength region.

In the work of Thorsbro et al. [6], the theoretical line formation of a spectral Sc I line was modeled to explore the effects of both temperature and HFS, using the BSYNand EQWIDTHcodes based on routines from the MARCSmodel atmosphere code [24]. One-dimensional (1D) MARCS atmosphere models were used. These models were hydrostatic model photospheres in spherical geometry, computed assuming local thermodynamic equilibrium (LTE), chemical equilibrium, homogeneity, and conservation of the total flux (radiative plus convective, the convective flux being computed using the mixing length recipe [25]). The resulting line strength measured in equivalent width is plotted against temperature in Figure 2. For the spectral line based on a single atomic level transition, we used the measured oscillator strength from Pehlivan et al. [11]. For the HFS based spectral line, we used the combined work of Pehlivan et al. [11], van Deelen [12].

From Figure 2, it is clear that for the cooler stars, it was crucial to have the correctly modeled HFS to explain the stronger Sc I lines. That a Sc I line has an HFS means that in reality, it consists of many weaker lines. That the many weak lines appear as a single line is due to the fact that the resolution of the spectrometer is not high enough to resolve all the details and also due to different line broadening effects caused by temperature, pressure, and other conditions in the observed star that make the weak lines blend together. The onset of saturation is delayed as the many weaker lines individually saturate later compared to that of a singular strong line.

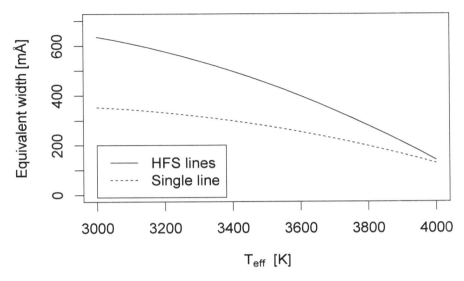

Figure 2. Figure from Thorsbro et al. [6] (their Figure 3). The equivalent width of a Sc I line as a function of temperature depends on whether the Sc I line is considered to be a combination of many small lines due to the hyperfine structure (HFS) or if the line is considered to be a singular atomic level transition. Notice how at high temperature, the two analyses converge, but at cooler temperatures, the HFS of the line enables the spectral line feature to become considerably stronger compared to basing the analysis on a singular atomic level transition. The metallicity and scandium abundance is assumed to be solar, and the surface gravity is assumed to follow isochrone relations with changing temperatures. Non-LTE and 3D effects are not considered.

2.3. Other Important Physical Effects

When it comes to accurate abundance analysis, having the correct atomic physics data is not always enough. Other effects need to be considered, like the 3D structure of the stellar atmosphere and its dynamics [26], as well as departure from local thermodynamic equilibrium [27,28]. This work recognizes that these effects are likely to be important for good abundance analysis of scandium in cool stars and therefore encourages this to be investigated further.

3. Conclusions

As shown in Figure 1, the strong scandium features found in stellar spectra dd not seem to be connected to the location of the observed star, as otherwise suggested by Do et al. [5]. Rather, it was shown by Thorsbro et al. [6] that an atomic physics approach was needed to explain the formation of strong lines, as shown in Figure 2. In particular, it was shown that the works of Pehlivan et al. [11], van Deelen [12] were needed to understand this line formation, showing that even today, there is still a need to investigate the atomic physics properties of many chemical species. The galactic center therefore did not seem to be that different from the solar neighborhood in this respect.

Funding: B.T. acknowledges support from the Swedish Research Council, VR (Project Number 621-2014-5640), Funds from Kungl. Fysiografiska Sällskapet i Lund (Stiftelsen Walter Gyllenbergs fond and Märta och Erik Holmbergs donation).

Acknowledgments: B.T. acknowledges the collaboration group through which the published work being reported on here was produced. The collaboration includes Nils Ryde, Michael Rich, Mathias Schulteis, Livia Origlia, Tobias Fritz, Henrik Hartman, Maria Lomaeva, Henrik Jönsson, Hampus Nilsson, Per Jönsson, Asli Pehlivan Rhodin, and Felix van Deelen.

Conflicts of Interest: The author declare no conflicts of interest.

Abbreviations

The following abbreviations are used in this manuscript:

HFS   Hyperfine structure
LTE   Local thermodynamic equilibrium

References

1. Bland-Hawthorn, J.; Gerhard, O. The Galaxy in Context: Structural, Kinematic, and Integrated Properties. *Annu. Rev. Astron. Astrophys.* **2016**, *54*, 529–596. [CrossRef]
2. Madau, P.; Dickinson, M. Cosmic Star-Formation History. *Annu. Rev. Astron. Astrophys.* **2014**, *52*, 415–486. [CrossRef]
3. Prialnik, D. *An Introduction to the Theory of Stellar Structure and Evolution*; Cambridge University Press: Cambridge, UK, 2000.
4. McWilliam, A. Abundance Ratios and Galactic Chemical Evolution. *Annu. Rev. Astron. Astrophys.* **1997**, *35*, 503–556. [CrossRef]
5. Do, T.; Kerzendorf, W.; Konopacky, Q.; Marcinik, J.M.; Ghez, A.; Lu, J.R.; Morris, M.R. Super-solar Metallicity Stars in the Galactic Center Nuclear Star Cluster: Unusual Sc, V, and Y Abundances. *Astrophys. J. Lett.* **2018**, *855*, L5. [CrossRef]
6. Thorsbro, B.; Ryde, N.; Schultheis, M.; Hartman, H.; Rich, R.M.; Lomaeva, M.; Origlia, L.; Jönsson, H. Evidence against Anomalous Compositions for Giants in the Galactic Nuclear Star Cluster. *Astrophys. J. Lett.* **2018**, *866*, 52. [CrossRef]
7. Ryde, N.; Fritz, T.K.; Rich, R.M.; Thorsbro, B.; Schultheis, M.; Origlia, L.; Chatzopoulos, S. Detailed Abundance Analysis of a Metal-poor Giant in the Galactic Center. *Astrophys. J. Lett.* **2016**, *831*, 40. [CrossRef]
8. Rich, R.M.; Ryde, N.; Thorsbro, B.; Fritz, T.K.; Schultheis, M.; Origlia, L.; Jönsson, H. Detailed Abundances for the Old Population near the Galactic Center. I. Metallicity Distribution of the Nuclear Star Cluster. *arXiv* **2017**, arXiv:1710.0847.
9. McLean, I.S. Science Highlights from 4 Years of NIRSPEC on the Keck II Telescope. In *High Resolution Infrared Spectroscopy in Astronomy*; Käufl, H.U., Siebenmorgen, R., Moorwood, A.F.M., Eds.; Springer: Berlin/Heidelberg, Germany, 2005; p. 25.
10. Nature, E. Nailing fingerprints in the stars. *Nature* **2013**, *503*, 437. [CrossRef]
11. Pehlivan, A.; Nilsson, H.; Hartman, H. Laboratory oscillator strengths of Sc i in the near-infrared region for astrophysical applications. *Astron. Astrophys.* **2015**, *582*, A98. [CrossRef]
12. Van Deelen, F. *Hyperfine Structure Measurements in Scandium for IR Spectroscopy*; Lund University Student Paper; Lund University: Lund, Sweden, 2017.
13. Marsden, G.C.; den Hartog, E.A.; Lawler, J.E.; Dakin, J.T.; Roberts, V.D. Radiative lifetimes of even- and odd-parity levels in Sc I and Sc II. *J. Opt. Soc. Am. B Opt. Phys.* **1988**, *5*, 606–613. [CrossRef]
14. Lawler, J.E.; Dakin, J.T. Absolute transition probabilities in Sc i and Sc ii. *J. Opt. Soc. Am. B Opt. Phys.* **1989**, *6*, 1457–1466. [CrossRef]
15. Öztürk, İ.K.; Çelik, G.; Gökçe, Y.; Atalay, B.; Güzelçimen, F.; Er, A.; Basar, G.; Kröger, S. Transition probabilities of neutral scandium. *Can. J. Phys.* **2014**, *92*, 1425–1429. [CrossRef]
16. Childs, W.J. Off-Diagonal Hyperfine Structure in Sc^{45}. *Phys. Rev. A Gen. Phys.* **1971**, *4*, 1767–1774. [CrossRef]
17. Zeiske, W.; Meisel, G.; Gebauer, H.; Hofer, B.; Ertmer, W. Hyperfine structure of CW dye laser populated high lying levels of ^{45}Sc by atomic-beam magnetic-resonance. *Phys. Lett. A* **1976**, *55*, 405–406. [CrossRef]
18. Ertmer, W.; Hofer, B. Zero-field hyperfine structure measurements of the metastable states $3 d^2 4 s^4 F_{3/2,9/2}$ of ^{45}Sc using laser-fluorescence atomic-beam-magnetic-resonance technique. *Z. Fur Phys. A Hadron. Nucl.* **1976**, *276*, 9–14. [CrossRef]

19. Siefart, E. Calculation of the Hyperfinestructure and gJ-Values of the 3d 4s 4p-Configuration of Scandium. *Ann. Der Phys.* **1977**, *489*, 286–294. [CrossRef]
20. Singh, R.; Rao, G.N.; Thareja, R.K. Laser optogalvanic spectroscopy of ScI: Hyperfine-structure studies. *J. Opt. Soc. Am. B Opt. Phys.* **1991**, *8*, 12–16. [CrossRef]
21. Aboussaïd, A.; Carleer, M.; Hurtmans, D.; Biémont, E.; Godefroid, M.R. Hyperfine structure of Sc I by infrared Fourier transform spectroscopy. *Phys. Scr.* **1996**, *53*, 28–32. [CrossRef]
22. Krzykowski, A.; Stefańska, D. Hyperfine structure measurements of the even electron levels in scandium atom. *J. Phys. B At. Mol. Phys.* **2008**, *41*, 055001. [CrossRef]
23. Basar, G.; Basar, G.; Günay, F.; Öztürk, İ.K.; Kröger, S. Experimental Investigation of the Hyperfine Structure and Theoretical Studies of the Even Configurations of Sc I. *Phys. Scr.* **2004**, *69*, 189. [CrossRef]
24. Gustafsson, B.; Edvardsson, B.; Eriksson, K.; Jørgensen, U.G.; Nordlund, Å.; Plez, B. A grid of MARCS model atmospheres for late-type stars. I. Methods and general properties. *Astron. Astrophys.* **2008**, *486*, 951–970. [CrossRef]
25. Henyey, L.; Vardya, M.S.; Bodenheimer, P. Studies in Stellar Evolution. III. The Calculation of Model Envelopes. *Astrophys. J.* **1965**, *142*, 841. [CrossRef]
26. Scott, P.; Asplund, M.; Grevesse, N.; Bergemann, M.; Sauval, A.J. The elemental composition of the Sun. II. The iron group elements Sc to Ni. *Astron. Astrophys.* **2015**, *573*, A26. [CrossRef]
27. Zhang, H.W.; Gehren, T.; Zhao, G. A non-local thermodynamic equilibrum study of scandium in the Sun. *Astron. Astrophys.* **2008**, *481*, 489–497. [CrossRef]
28. Bergemann, M.; Nordlander, T., NLTE Radiative Transfer in Cool Stars. In *Determination of Atmospheric Parameters of B-, A-, F- and G-Type Stars*; GeoPlanet: Earth and Planetary Sciences Series; Niemczura, E., Smalley, B., Pych, W., Eds.; Springer International Publishing: Cham, Switzerland, 2014; pp. 169–185. ISBN 978-3-319-06955-5. [CrossRef]

MDPI
St. Alban-Anlage 66
4052 Basel
Switzerland
Tel. +41 61 683 77 34
Fax +41 61 302 89 18
www.mdpi.com

Atoms Editorial Office
E-mail: atoms@mdpi.com
www.mdpi.com/journal/atoms

Lightning Source UK Ltd.
Milton Keynes UK
UKHW051906091020
371310UK00007B/136